Military Brass vs. Civilian Academics at the National War College

A Clash of Cultures

Military Brass vs. Civilian Academics at the National War College

A Clash of Cultures

Howard J. Wiarda
Dean Rusk Professor of International Relations
University of Georgia

Senior Scholar
Center for Strategic and International Affairs (CSIS)
Washington, D.C.
and
Public Policy Scholar
Woodrow Wilson International Center for Scholars
Washington, D.C.

LEXINGTON BOOKS

A division of

ROWMAN & LITTLEFIELD PUBLISHERS, INC.
Lanham • New York • Toronto • Plymouth, UK

Published by Lexington Books
A division of Rowman & Littlefield Publishers, Inc.
A wholly owned subsidiary of The Rowman & Littlefield Publishing Group, Inc.
4501 Forbes Boulevard, Suite 200, Lanham, Maryland 20706
www.lexingtonbooks.com

Estover Road, Plymouth PL6 7PY, United Kingdom

British Library Cataloguing in Publication Information Available

Library of Congress Cataloging-in-Publication Data

Wiarda, Howard J., 1939–
Military brass vs. civilian academics at the National War College : a clash of cultures / Howard J. Wiarda.
p. cm.
ISBN 978-0-7391-5085-6 (cloth : alk. paper)—ISBN 978-0-7391-5087-0 (electronic)
1. National War College (U.S.) 2. Military education—United States. 3. Sociology, Military—United States. 4. Academic freedom—United States. I. Title.
U412.W53 2011
355.0071'173—dc22
2010048619

∞™ The paper used in this publication meets the minimum requirements of American National Standard for Information Sciences—Permanence of Paper for Printed Library Materials, ANSI/NISO Z39.48-1992.

Printed in the United States of America

Contents

Preface

In January 1990 my wife Iêda Siquiera Wiarda, a fellow political scientist, accepted a new position as the Portugal-Brazil specialist at the Library of Congress in Washington, D.C. At the time we were living in Amherst, Massachusetts, where both of us were teaching at the University of Massachusetts, so the new job would require Iêda moving to Washington. Further complicating the situation was the fact we had a teen-aged son, a junior in high school, doing well, with nice friends, and we wanted Jonathan to finish high school with his class.

What to do? Would we all move to Washington? Would I give up my tenured full professorship at UMass? Would we pull Jonathan out of school and enroll him at a new high school in or around Washington where he had no friends? Or, as we'd done before, not always the best arrangement, would we opt for the life of a commuter marriage, with one part of the family in Amherst and the other part in Washington?

I was the one who made the decision. For years Iêda had been sacrificing her professional career for the sake of advancing mine. The Library of Congress position was the opportunity of a lifetime for her. It was a great job and it exactly fitted her research specialization. And, to be honest (she doesn't know this), I still felt badly about a family decision we'd made some years earlier, involving a similarly attractive job offer for her but again in another city, which she'd reluctantly turned down for the sake of keeping our family and children, then much younger, together. But now I told Iêda to accept the Library offer. I would stay back in Amherst to take care of Jonathan, and we would commute on weekends.

So that's how I got, eventually, to the National War College (NWC) in Washington, D.C. With Iêda thriving in her LC position, and Jonathan in his senior year of high school and preparing to go to a university, I began to search for a position in D.C. to bring our family back together again and so we wouldn't have to do all that commuting. I tried several possibilities: George Washington University, American University, George Mason University, and Georgetown.

At GW I had friends in both the political science department and the Elliott School of International Affairs; the problem was they had no positions available and no money. At Georgetown the situation was similar: they said they'd love to have me on the faculty but they had no money or positions available. At AU, where I was a finalist candidate for the department chairmanship, the problem was the dean of the foreign service school; he wanted another person besides me in the chairmanship. At George Mason the problem was similar: an uncertain program in a still uncertain school that was not sure in which direction it wished to go.

And then I heard about a new position open at the National War College, at Fort McNair, in Washington, D.C. I applied for, went through the interview process, and got the job offer at NWC. Although there were still several possibilities out there in the bush, this was the only bird that I actually had in hand. I visited NWC, discovered they were overjoyed to have me, and accepted their offer. But to hedge my bets, I only took a leave of absence from my tenured position at UMass rather than resigning the position.

I had little idea what to expect at NWC. Although I'd previously lectured at all the military academies, I'd never held a position at any of the service schools. Nor did I come from a military family. My father had fought in World War I, and three of my uncles fought in World War II, but none of them had made a career out of the military. So *everything* at NWC was going to be new to me. And let me emphasize right at the beginning that NWC is preeminently a *military* institution (the National *War* College) much more than it is a civilian college or university.

And that is the tale told in this book. NWC is the story of two cultures, two entirely different ways of life, coexisting uncomfortably and maybe incompatibly within this single institution. On the one hand, NWC presumes to be a "college," within a larger "university" (the National Defense University), with all that is assumed within those terms—academic freedom, vigorous research activities, free speech, tenure, freedom of activity within the classroom, encouragement of research and publication, a merit system, professional ranks, and so on. But in fact, none of these features exists within NWC.

Instead, it is the military and militaristic aspects of NWC that predominate: the command structure, the system of ranks and hierarchy, a system of top-down authority, a system of orders emanating from above and of obedience expected at the bottom, a system that values students more than faculty, views the faculty (like military officers) as interchangeable pegs in a well-oiled machine, sees no value in research, tenure, or individual expertise. In short, NWC is an Army base, not a serious academic institution.

These two systems and the cultures that go with them—military, on the one hand, and civilian academic, on the other—coexist but barely so at NWC. The situation is explosive or nearly that. It is like the "two cultures" made famous in the novels of British writer C. P. Snow. One group doesn't understand, comes from a different sociological and political background, and harbors resentments toward the other. Those resentments are reciprocated by the other side. The two cultures operate on different assumptions and function on different wavelengths. There are simmering tensions between them which flare to the surface on an everyday basis. The two cultures work at cross purposes; they exist in a state of endemic civil war and rebellion.

This is the story of these two cultures, as they interact at the National War College (NWC). NWC is the nation's preeminent and highest-level institution for advanced military training in war, politics, and international affairs. Former National Security Adviser Brent Scowcroft and former NSA *and* Secretary of State Colin Powell, among many other high officials, all trained there. So what goes on at NWC has vast importance for the nation's military, strategic, and international posture. And if the National War College is as deeply divided and dysfunctional as we portray it here, that is a cause for major national concern.

The book that follows is written in part in the first person. It is a personal memoir as well as a deep probing into the inner workings of the National War College. It is a serious study but it is also based on my own experiences working and teaching at NWC. During that time I experienced the War College at all its levels: the president's office, the commandant, several deans, the core course, so-called "advanced" seminars, regional studies, the Reserve Officers' Training Component, military bases at home and abroad, the academies, the other service-level war colleges. I've seen it all—often more than I ought to have seen. This is my accounting of what I've seen and experienced—and, by extension, why, in part, we're in such bad trouble as a country.

In writing this book I am indebted to many persons. First and foremost is my wife, Dr. Iêda Siqueira Wiarda, who shared both the glory and the agony of my stint at the National War College. Friends and colleagues Dale Herspring and Roy Stafford read and commented on an earlier draft of the manuscript. Annie Kryzanek has been my indispensible research assistant, while Kathryn Johnson and Doris Holden, my long-time friend, editor, and collaborator, worked on the manuscript. My greatest debt, however, is to the colleagues, faculty, and students at the National War College. They are a quite remarkable bunch. I would love to name them all individually but, first of all, they are too numerous to name and, second, by doing so it would surely give away who talked to me about this book, who provided me with necessary information, whose conversations I have reproduced here. Since I don't want to add any more to the incessant behind-the-back conversations (and frequent back-stabbing) that go on at NWC, I prefer to keep those names private. That choice, however, means that the usual author disclaimer that comes at the end of a preface applies in spades here: I alone am responsible for the analysis and views that follow.

Howard J. Wiarda
Tower Villas
Arlington, Virginia
Fall, 2010

Chapter 1

Geography and History

GEOGRAPHY

The National Defense University, which is the umbrella organization that houses the National War College, is situated at the historic southern "entrance" to Washington, D.C. It is located in southwest Washington at the point (Turkey Buzzards' Point) where the Anacostia and Potomac Rivers come together. The War College itself, occupying (Teddy) Roosevelt Hall, is right at the confluence of the two rivers; it is one of the main buildings in a complex situated at Fort McNair, which was originally founded as a military base to defend Washington from incursions up the Potomac. If you're sailing north up the Potomac from Chesapeake Bay or landing from the south at Washington's Reagan National Airport, you sail past and can easily see the War College grounds off the right side. Of course, the original defensive purposes of Fort McNair to guard the capital are no longer relevant; hence, the location in this historic site of the National War College.

The neighborhood where the base is located is not the greatest. Southwest Washington is an area of severe contrasts. On the one side are the plush townhouses and luxurious high-rise apartments, called "Tiber Island," that line the Potomac and the opening to the Washington Channel. But within a short couple of blocks one finds slums that are as bad as any in Washington. M Street Southwest, the main thoroughfare through the area, has a Giant grocery store in a shopping center and is undergoing renovation; but only a block from that one finds crime, drugs, and high unemployment. The neighborhood includes the Maine Street fish market, luxury hotels along the waterfront, and the Arena Theater, but also sleazy nightclubs, junk yards, and construction and trucking companies whose soil beneath reeks of contaminants. The new Washington Nationals' baseball stadium was supposed to spark a gentrification and revival in these blighted neighborhoods, and it has done that, but so far not nearly to the extent expected.

It's sometimes dangerous to drive through the neighborhood, let alone have your car get stalled there. You don't want to be there at night; I knew military officers at NWC, both students and faculty, who would not allow their wives to drive to the base alone.

But once you drive through the gates of Fort McNair (after 9/11 security became much tighter and it became difficult for unauthorized civilians in taxis or private cars to enter the grounds) a different world opens up. In contrast to the grime and litter outside the gates, NDU is spotless. The chaos, crime, and confusion of a blighted urban neighborhood are replaced by military order, cleanliness, and discipline. The streets on the outside are strewn with broken glass, drug syringes, and the occasional condom; on the inside, there's never a speck of trash. The outside looks like a concrete jungle; on the inside, it's lush, green, and carefully manicured—especially the nine-hole golf course that, in my time there, dominated the central commons/parade grounds. For a long time the course provided a hint of NWC's priorities: one time I arrived on "campus" when the nearby rivers were rising and noted that enlisted men were sandbagging the golf course but not the academic buildings. Now, the golf course is closed, a victim of administrative decisions and Defense Department budget cuts.

As you pass through the main gate, the military police (now this function has been taken over by less-smart private security) check your identification. I had a Department of Defense, Fort McNair military sticker on my car, so I breezed through easily, but visitors had to show their driver's license and state their purpose in visiting. Post-9/11, you need, in addition to showing an ID, to get out of your car, open the hood and trunk for a search, open all your car doors for an additional search, and be subjected to a long-handled mirror inspection of your car's underside.

On the base the speed limit is twenty miles per hour, and it is enforced! I find it impossible to drive that slowly and have been the recipient of numerous speeding tickets. The Military Police are ruthless and unforgiving in their pursuit of speeders. One of my observations is that military officers driving on the base find it somehow easier to obey that speed limit (it's the military requirement that you obey orders) than do civilians (undisciplined, those slackers!). One of my civilian academic colleagues, European politics specialist Hugh De Santis, who got many more speeding tickets even than I, had the embarrassing and demeaning experience of having the MPs come to his office, hoist him up under both arms, and frog-walk him away to their superiors for further questioning and a reprimand. Hugh was actually put on probation by the War College because of his driving habits!

Passing through the gate, there are historic, nineteenth-century barracks and offices on both sides that go on for a block. On the far right-hand corner is the Inter-American Defense College (IADC), a kind of lower-level war college for officers from all over Latin America and where I have lectured on numerous occasions. It remains questionable just how much the IADC has succeeded in changing the behavior of the Latin American officer corps. The Latin American officers and their school are looked down on and frequently treated with condescension by the American officers and the other colleges on the base.

One block from the entrance the road comes to a "T." Side roads lead to the right (west, toward the Potomac) and left (east, toward the Anacostia). In front of you is a vast expanse of well-trimmed green grass, first the military parade grounds, then the tennis courts, then the golf course mentioned earlier—all this stretching beautifully for a full verdant half-mile before terminating at the impressive building, with its great and fierce eagle out front, that is the War College.

Parallel roads passing north and south flank this great green expanse, which clearly is the most prominent feature of Fort McNair, running along both sides of the base, joining again at the far end in a loop around both the front and back of the War College building. Along the right-hand side is "Generals' Row" (you must be at least a two-star to live there), with beautiful, stately, two-story brick colonial homes that back onto the Washington Channel, a branch of the Potomac. Midway on this route, near the tennis courts and golf club, is the Fort McNair Officers' Club, the best bargain in the city where you can enjoy the elegant dining room or sit out overlooking the water (or the golf course), enjoy the always-good (a combination of *haute cuisine* and southern fried), all-you-can-eat buffet, and unlimited quantities of wine—and all this for the price then of $6.95—a wonderful place to take your wife, a date, or your secretary.

In the middle are two of the historic buildings on the base, used as administrative offices and sometimes to house foreign military officers as part of the NDU program to reach out to old and new allies. Just east of these buildings and backing against the far wall of the base is the new, modern Marshall Hall, the administrative headquarters of NDU containing the offices of the president, the library, and conference and seminar rooms. Just north of that is the brand new Lincoln Hall, which houses the Institute for National Strategic Studies (INSS) and the Center for Hemispheric Defense Studies (CHDS) where I spent part of my time. Just to the south is Eisenhower Hall, a '50s era building that is home to NWC's sister institution and sometimes rival, the Industrial College of the Armed Forces (ICAF), founded in 1924, which deals with such issues as procurement of military supplies, logistics, transport, etc. Because of its focus on supplies and procurement, ICAF was sometimes known as "Black and Decker University"—after the hardware and tool manufacturing company. In contrast, NWC, where I taught, focuses mainly on international policy and strategy.

The main road around the base or campus (depending on whether you're military or civilian) is one-way and the traffic flow is counter-clockwise. So after you leave that first T, you go right or west for a block, then turn south along Generals' Row, proceed past the tennis courts and Officers' Club, and then continue another quarter mile south through the old golf course and on to the War College whose majestic building dominates the base at the confluence of the two rivers. Looping around NWC, you would then proceed north, past Eisenhower, Marshall, and Lincoln Halls, and eventually back to the T and then the gate.

Roosevelt Hall, at the fulcrum of Fort McNair, is a majestic building that dominates and is the focal point of the entire grounds. Plans for the building were drafted in 1901 when Elihu Root was Secretary of War and Teddy Roosevelt was president. The same trowel that George Washington used to lay the cornerstone of the Capitol was used for the cornerstone of the new war college, a ceremony that Root, President Roosevelt, and Secretary of State John Hay all attended. The building was designed by the architectural firm of McKim, Mead, and White, built by the Army Corps of Engineers for a modest $700,000, and dedicated in 1908.

The style of the architecture has been described as "Roman basilica." The building is 250 feet wide and 125 feet deep. It is of red brick with limestone ornamentation and massive soaring arches. The roof is dark slate, meant to last. It is a beautiful building, classic in design and nicely proportioned. At the entrance is a fierce American eagle. Upon completion of the building, the German government donated a statue of Prussian emperor Frederick the Great that also stood in the front entry, but two world wars in the twentieth century waged against Germany resulted in Frederick being permanently removed.

Passing through the front doors one enters a massive, beautifully designed, three-story atrium with railings on all four sides on the second and third floors that is an engineering marvel. Parties, dances, receptions, and other "high level" social events are occasionally held here. The first right-hand hallway leading off the atrium is the "State Department Corridor," culminating in the office of the Foreign Affairs Deputy, which is also the office where George Kennan, the first Foreign Affairs Deputy, wrote his famous "Mr. X" article outlining the threat from the Soviet Union and setting forth the "containment" strategy for countering it. The second right-hand hallway houses the College's Executive Officer (XO), who doubles as the dean of students, in the past a military officer but now a civilian (retired military), and the office of the Commandant, ordinarily a two-star general, which looks out on the two rivers as well as Reagan Airport. In between these two corridors is the old library, now moved to Marshall Hall, the far end of which is used for ceremonial events (promotions, going-away parties) and overlooking the Channel. Down the parallel left-hand corridors of the main floor are faculty offices, classrooms, administrative offices, and the office of the dean of faculty. In the middle, off the atrium, is the large auditorium, Arnold Auditorium, into which an entire NWC class can fit.

The second floor, on both east and west sides of that beautiful, three-story atrium, houses additional faculty offices and seminar-size (space for fifteen to twenty persons) classrooms. Each individual service, including now the Coast Guard, has its own corridor here, decorated with heroic service pictures, plaques, and insignia. Recent renovations have made the faculty offices brighter and more commodious; the classrooms have also been redecorated.

On the third floor, under the eaves, which used to be mainly a storage and throw-away area (but where I kept a private quiet, hide-away office to do my writing—a no-no at the War College; you're not supposed to "hide" from your students), the usable space has now also been converted into offices. One additional feature of the third floor: it houses the spectacular, birds' nest "Capstone" Seminar Room which, again, overlooks the joining of the two rivers and offers a magnificent view to the south; Capstone Courses (for senior or "flag" officers—generals and admirals) and other special seminars are held here.

The students are housed in the basement, at ground or parking-lot levels. The student body when I was there numbered 180 (now it is over 225), which is then subdivided into smaller groups. Each student group of fifteen to twenty, containing a mix of the services plus civilian and international students, has its own room (usually complete with a coffee machine and a small fridge) and each student has his own desk or carrel. The students do not live on the base; they live at home or must find their own accommodations, but they are given desk space for books and personal items. The basement also houses student mailboxes, an informal lounge area, and a well-equipped workout room.

HISTORY

As early as colonial times, the strategic value of what was then called Turkey Buzzards' Point was recognized. Anyone coming up the Potomac River would immediately recognize that the Point commanded the entry to both the Anacostia River and the wider stretches of the Potomac itself just before Washington. As early as 1693 there is mention of a stockade fort at the Point.

Like so many U.S. cities of that early time, Washington, D.C., was founded where two bodies of water, in this case two rivers, came together.

During the colonial period, this stretch of the upper Potomac was only thinly settled. Most of the early colonists, including the ill-fated Jamestown settlement, were located farther downstream on the river, on tributaries, or around the Chesapeake Bay. The "fort," so-called, was often deserted and fell into disuse.

However, once the United States became independent, wrote a new constitution, and definitively moved the capital from Philadelphia to Washington, D.C., Buzzards' Point took on additional, or renewed, strategic significance. Take a look at Peter Charles L'Enfant's original, 1791 plan for the capital: he has it nestled neatly and about equidistant between the two rivers. Anyone with any strategic sense could see that the command of the point where the two rivers joined was essential to the southern defense of the city. An enemy sailing up the Potomac or into the Anacostia would have to get past these fortifications to attack the city (see Figure 1.1).

What would later become Fort McNair clearly shows on L'Enfant's city design for the new capital. At that time, the Anacostia River was still referred to as the "Eastern Branch" (of the Potomac). The area where the Eastern Branch flowed into the Potomac was then called Carrolsburg. It was a small settlement set at right angles to St. James Creek—now the St. James Marina, just outside Fort McNair's eastern gate. Carrolsburg was located just east of the Creek at the spot where the building housing the U.S. Coast Guard now stands.

The other small settlement in the area at this time, called Hamburgh, was on the Potomac River itself, just south of Georgetown and about where the Kennedy Center and the Watergate Apartment complex now are. Once the decision had been made to place the new capital at this location at the confluence of the two rivers, Carrolsburg and Hamburgh competed to have it located closest to themselves. But L'Enfant, backed by his collaborator in this project George Washington himself, had a bolder and bigger vision of the new capital: he located it between the two rivers and at the same time encompassing the new capital. If one thinks of the capital as an inverted isosceles triangle, then the Capitol and Capitol Hill form one, the eastern, point of the triangle; the White House, Foggy Bottom, Hamburgh, and now Georgetown encompass the western point of the triangle; and Carrolsburg and the fortress at Buzzards' Point constitute the third or southern point of the triangle. The Fort would guard both the Potomac and the Anacostia approaches to the city. See the attached map from L'Enfant's original plan.[1]

However, in the fort's first big test as a guardian of the new capital, it didn't quite work out as intended. In the War of 1812, the British did not risk coming up the Potomac and having their ships sail under the fort's guns. Instead, they sailed

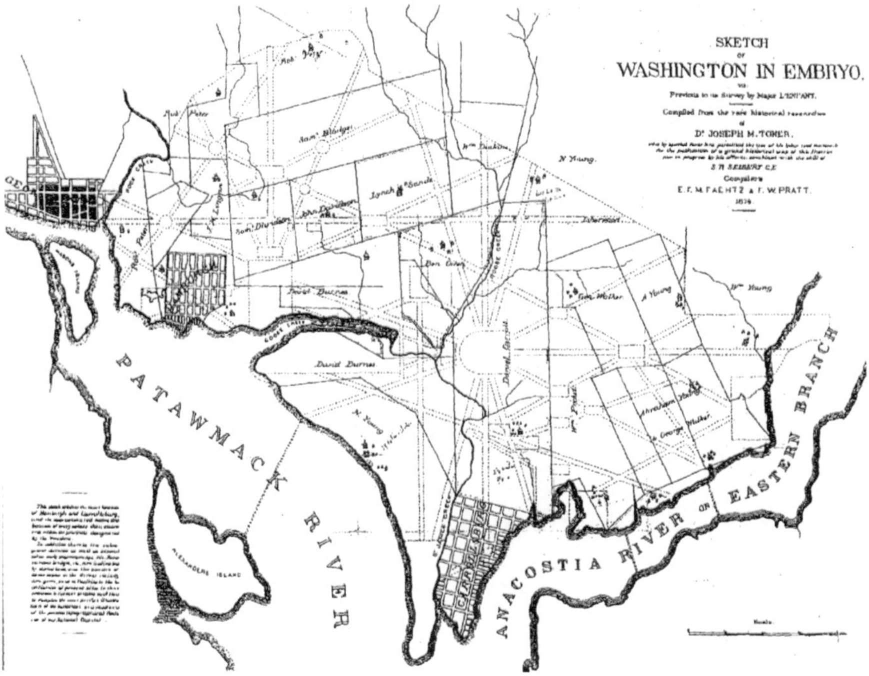

Figure 1.1.

up the Chesapeake Bay, parked their ships, and attacked Washington overland by marching across southern Maryland. This is the famous episode where the British troops attacked and set fire to the White House, forcing President James Madison to flee northward up to the small town of Brookeville, Maryland, twenty-five miles away.[2] Of course, the British, a few days later, retreated the same way, again bypassing the guns at the Buzzards' Point fortress.

Its worthlessness as a defense of the capital city having been dramatically demonstrated, the fort, during most of the nineteenth century, fell into disrepair and disuse. It was used mainly as a weapons storage facility or arms depot, and not as a serious fortification or line of defense. That was reflected in the several name changes that it went through in the nineteenth century, from the Fort at Turkey Buzzards' Point to the United States Arsenal to the Washington Arsenal to, in 1881, the Washington Barracks. It was only in 1948 that it was renamed Fort Lesley J. McNair.

After nearly a half century hiatus, the fort's usefulness in the defense of Washington was recognized again during the Civil War. It continued to be used as an arms storage depot, but its fortifications were reinforced and updated. Fortunately for Washington, the armies of the confederacy never attempted an up-the-Potomac assault on the capital that would have resulted in a direct confrontation with the Union garrison stationed at Buzzards' Point because by then modern armaments, especially cannons, had rendered the ancient fort all but indefensible.

The fort, known then as the Washington Arsenal, did achieve considerable notoriety as the war drew to a close. First, the trial of the Lincoln Conspirators, four persons engaged in a plot to kill the president, was held in 1865 in one of the oldest buildings of the complex. Convicted, the four were then hanged in the yard of the penitentiary in the center of the historic parade grounds, and buried nearby, only a few steps away from what is today the base tennis courts. One of the four was a woman, Mary Surratt, who continued to proclaim her innocence until her death. It is said that, when it snows in Washington, D.C., the snow never stays on the ground over Mary's burial spot because she continues to churn in her grave. The fact that the base's steam heating pipes pass under this spot may provide an alternative explanation.

John Wilkes Booth, the Lincoln assassin, was similarly brought to the Arsenal for burial in 1865. He was buried under one of the cells of the penitentiary, but the body was later transported to a family burial ground in Baltimore.

After the Civil War, the fort again fell into disuse. It was not deemed critical for the defense of Washington and no one quite knew what to do with it. Besides, both the Union Army and the Army of the Confederacy had been disbanded; we had no standing army to maintain and occupy it. It continued to be used as an arms storage depot, but by the 1880s its mission had changed. New buildings (still standing and in use) were put up and its name changed to that of Washington Barracks. As the United States began to emerge during this period of the late nineteenth century as a would-be global power, a contingent of troops was kept at the fort and new uses were contemplated for the grounds.

In 1901, fittingly during the expansionist and adventurous Teddy Roosevelt Administration, plans were drawn up for the construction of the beautiful National War College building at the point of the two rivers, as we know it today. In those days known as the Army War College, the building was dedicated in 1907 and the first class graduated in 1908. It is not a coincidence that the War College was planned and built precisely at the time the U.S. was emerging as a world power with global ambitions for the first time. For in 1898 we had defeated Spain in the Spanish-American War, acquired the Philippines, Puerto Rico, the Panama Canal Zone, and (for a short time) Cuba as colonies, were occupying Haiti, Panama, Nicaragua, the Dominican Republic, and Cuba off and on, and were establishing a string of bases in the Pacific (Hawaii) and throughout the Caribbean stretching from Florida to Venezuela.

During World War I, some of the teaching and training functions of the War College were suspended, in the interests of the greater war effort. They were resumed after the war; but, as usual following U.S. wars, the country had demobilized, we had returned to "normalcy"; and we did not keep up a large standing army. In World War II classes were again suspended and the War College closed down as the base was converted into the Army General Headquarters in 1940. Its location only a mile from the White House and the next-door Army War Department (at that time housed in what is today the Executive Office Building on the White House grounds) made this a convenient switch.

The idea for a *National* War College, as distinct from the old *Army* War College, came out of the American experience during World War II. The inspiration was the same one that led to the creation in 1947 of the National Security Council; indeed, the creation of the War College was part and parcel of, and emerged from the same experiences as, the massive defense reorganization of 1947 (The National Security Act) that created the Joint Chiefs of Staff (JCS), the Office of the Secretary of Defense (OSD), and the new Department of Defense. The basic problem, which the war brought dramatically to the surface, was a lack of coordination between the three main services, which were often fighting three different wars, and between the military services and such civilian agencies as the State Department.

The creation of the National War College was meant to correct these deficiencies. The rationale was the same as for the creation of the NSC. During the war, President Franklin Roosevelt would get lengthy analyses, often obscure and full of special pleading, from the Department of the Army, the Department of the Navy, the State Department, and other agencies. He needed someone or an office in the White House to boil these analyses down, remove the misinformation and special pleading, and present the information to him in an organized way that facilitated presidential decision-making.

Much the same logic applied to the founding to the War College in 1946. During the war, the American Joint Planning Staff was coordinated with a comparable British group to comprise the Combined Planning Staff that supported the combined Chiefs of Staff. But repeatedly during the war, the Americans discovered that British members of the Combined Staff presented a unified position in a single memorandum on military plans and operations, while the American "position" was represented by three separate and uncoordinated memoranda: one each from the Army, Navy, and Air Force. The Americans then had to negotiate, often at length, among the three services to arrive at a single, unified position at which their British counterparts had already arrived. Hence, the idea was for a single, joint war college incorporating all the services and leading, presumably, to a common position on defense strategies and issues.

Planning for the new war college began in 1942–43, in the middle of the war. The first results were not very promising: what emerged in 1943 was a much narrower proposal for an Army and Navy Staff College (ANSCOL), involving a four-month course, in which Army officers would, incongruously, be trained at the coastal Naval War College in Newport, Rhode Island; Navy officers, at the landlocked Army base at Fort Leavenworth, Kansas; and Air Force personnel, in Florida. All the officers would then meet together in Washington later in the year. This unwieldy plan led, arguably, to even less coordination among the services and less "jointness" than had existed before. Initially, to speak to the second problem identified earlier, no State Department or other civilian agency personnel were included in the program.[3]

The issue of higher military education, it was clear, could not be separated from the larger question of military reorganization that was occurring in the post-war period, and particularly as the Cold War began to heat up in 1946–47. Several internal reports reflected the need already seen during the war for a high-level and, above all, *joint* war college. The driving forces behind these proposals were mainly Lieutenant Generals Henry Arnold (after whom Arnold Auditorium in Roosevelt Hall is named) and John DeWitt, supported at the top by Generals George Marshall and Dwight Eisenhower. The Joint Chiefs had instructed General DeWitt, then serving as commandant of ANSCOL, to prepare a plan for the creation of such a joint college with the following objectives:

1. To promote among the services knowledge of each other.
2. To promote teamwork among the services.
3. To prepare officers for joint operations.
4. To prepare officers for command of joint operations.
5. To prepare senior officers for high-level, Defense Department staff functions.
6. To promote understanding between the services and other civilian branches.

The plan De Witt submitted was highly controversial. The three services were not at all agreed on the necessity of joint operations or of subordinating their particular service interests to a new joint command. Especially the three already existing war colleges—Army, Navy, and Air Forces—saw no need for a new institution to which they would, presumably, play second fiddle. The Army War College, for instance, already did the training for the Army Air Force as well as having some Navy officers on its staff, so it argued *it* should be the venue at which joint training took place. Meanwhile, some Navy admirals tried to scuttle the plan politically through their allies in Congress.

In December 1945, President Truman sent Congress a "special message" recommending the creation of a single, joint Department of Defense. The fact that the proposal for a "Joint College," which was to be the National War College, was put forward the very next month, in January 1946, indicates how closely overall military reorganization was related to reform of military education. The "Gerow Board," which met and planned the war college during this period, was remarkably prescient in its recommendations: it called for not only a National War College but also a "National Security University" (not implemented until 1975) as an umbrella organization to oversee the College's as well as other programs, a reorganized Industrial College of the Armed Forces (ICAF) that would provide training in procurement and industrial mobilization, and even a State Department College—the forerunner of the Foreign Service Institute (FSI).

The National War College, according to its founding documents, was to be "concerned with grand strategy and the utilization of the national resources necessary to implement that strategy." The planned course of instruction was to include:

1. Grand strategy and war planning.
2. Foreign and domestic policies of all nations.
3. Causes and prevention of war.
4. Economic and social resources of nations and their relations to war potential.
5. Joint policies and doctrines.
6. Mobilization and demobilization (really an ICAF function).
7. Policies for cooperation with allies.
8. Trends of future wars and their implications.
9. Armed forces' responsibilities after cessation of active hostilities.

As revealed in February 1946, the plans for the War College included State Department and other civilian agency participation, operations and training conducted under the authority of the Joint Chiefs, and a ten-month course beginning in September of that same year. The new College was to be located in Washington, in Roosevelt Hall, on the site of the previous Army War College, which soon moved to Carlyle Barracks, Pennsylvania. There was still lingering dispute over the name of the College, with some preferring "College of National Defense" and others, "College of National Security"; but in the end, "The National War College" was decided on by Secretary of War Robert Patterson.

By March 1946, a committee of academic experts, including Arnold Wolfers of Yale, Walter Wright of Princeton, William Langer of Harvard, Calvin Hoover of Duke, Edward Earle also of Princeton, and Janice Baxter, president of Williams College, had been convened to design the new college's curriculum. They recommended "unanimously" a comprehensive lecture program with a seminar and small-group method of instruction. They also recommended that the foreign policy/international relations part of the curriculum be offered in the fall term so that leading professors could more easily go on leave from their home universities. "War studies" were to be offered in the spring.

The War College was located on an Army base; it had a top-down military command structure; and its commandants were to be military generals or admirals. The War College is enormously proud of the fact, however, that the well-known diplomat and writer George F. Kennan was recruited to be the first civilian Foreign Affairs Deputy for "matters pertaining to international relations and foreign policy"; but Kennan was only an assistant to the top brass on issues pertaining to the course of study. It is very clear from the beginning that, even with some limited civilian input and some civilian students (now approximately 25 percent), the War College was to be preeminently a military school and not a civilian university, with the Army firmly in control, and with a hierarchical command in which civilian instructors take orders from military superiors.

The size of the student body was initially set at one hundred members, growing over time to 180 and now 230–40. The Army and Navy occupied a disproportionate share of the places available; the Air Force was underrepresented; and at first there was only a token handful of civilians. The faculty consisted of eighteen military officers, mostly recruited from the old ANSCOL, and a slightly smaller group of civilians, from ANSCOL as well as the recently disbanded Army War College. George Kennan was the most distinguished civilian faculty member, but Arnold Wolfers and Bernard Brodié added further distinction. Kennan was "dual hatted," serving as Foreign Affairs Deputy to the Army commandant and head of the civilian foreign affairs faculty. He and others helped establish the year-long course curriculum that, not very different from today's program, consisted of five segments: international affairs, intelligence, logistics, strategic operations, and tactical operations. The first class in the new college matriculated in the late summer of 1946.

Kennan was by far the best-known and most dynamic leader of the War College in its first years. He not only wrote his famous "Mr. X" article while ensconced in his corner office at the NWC; he also designed and wrote the foreign affairs curriculum and delivered many of the lectures in the first segment of the course. Not well known is the fact that Kennan pulled some of these early writings and lectures together into a volume entitled "Measures Short of War," published by the NDU Press. I have one of the few copies available in my personal library. In this book Kennan wrestled with issues of diplomacy, what we would today call "soft power," how military might can be used to reinforce diplomacy, and how diplomacy can support the use of force. It is a fascinating book full of insights for today's policy-makers.

Most interesting in terms of the themes of this book is what Kennan had to say about the War College itself. He envisioned it as standing at the apex of the Professional Military Education (PME) system in the United States, above the separate service war colleges. He saw it as both a joint institution among the services, and a way of integrating civilian and military services. Kennan viewed it as essential for State Department officials to learn about the military instrument and for military officers to know the techniques and utility of diplomacy. Combining war-fighting and diplomacy, Kennan envisioned a school that would train the future Brent Scowcrofts, Colin Powells, and James Joneses of this world.

He also envisioned a college that would be in the forefront of international affairs research and innovative strategic thinking. Not content with rote learning, memorization, and teaching from prescribed texts, Kennan wanted the War College faculty itself to do research, publish, and be on the cutting edge of new ideas in the then-new, mushrooming field of international relations. While Kennan and Brodie were there, NWC lived up to these ideals; after they left, the military side of the College came to dominate and the civilian academic side was downgraded. New, original, and innovative thinking and research were not only not done, they were positively discouraged by the military commanders. They believed not only in "going by the book" but also in teaching from the book—i.e., from prescribed texts, which did not include the writings of their own faculty. As we see in considerable detail later on, that continued to be a serious problem when I joined the War College faculty during the 1990s.

After Kennan left the War College, the quality of its civilian faculty declined considerably. In part, that was due to the difficulty of getting quality faculty to take a four-month leave from their universities, at very modest pay, with a heavy teaching load, and with very few research opportunities. At this time, the College had no full-time or permanent civilian faculty, a problem that continued to plague the College in my time. In part also, it had to do with the fact that the military side, the three services but particularly the Army, wanted to control and dominate the College.[4] Military officials tended not only to resent the civilian academics and their often free-wheeling ways, they also did not like the research orientation and policy soul-searching that Kennan had introduced. To their way of thinking, that introduced too much discord and questioning into the program, and it might conflict with official

DOD policy. They much preferred the top-down curriculum of the staff-college model.

The result was that, over the next years and decades, the College went into a long-term decline. The civilian faculty was mediocre and the military faculty favored a from-the-book approach. The thinking, research, and writing side of the program that Kennan had championed was largely abandoned in favor of a hard-line, unimaginative, Cold War orthodoxy. The College was hardly the place for the speculative and imaginative thinking that Kennan had wanted. A large plaque outside Kennan's former office justly celebrates his famed "Mr. X" article—arguably the most influential foreign policy essay of the last sixty years—but completely neglects to mention the cultivation of an intellectual climate at the College that would make that kind of intellectual creativity possible.

Periodically in the '60s and '70s, the sad state of the War College would be lamented and complained about, and the Congress would launch an investigation. Not much came of these reviews, however, except modest reforms; it seemed the College always reverted to form, which was: military-dominated, a weak faculty, no or little research agenda, not a very exciting place to be. The College had lost its place at the "apex" of the PME system; indeed, that designation, which was clearly the intention of NWC's founders, both military and civilian, was never officially conferred. Meanwhile, the Army War College, the Naval War College, and the Air War College, as well as ICAF, which under the early plans were supposed to be subordinated to NWC, reasserted themselves and claimed equality, if not superiority, to the supposedly *National* War College. It was another instance in a long tradition of these of the separate services reclaiming their positions and prestige over any and all "jointness" efforts.

A fundamental reorganization occurred in the mid-1970s: to this point the National War College had been the only occupant of Roosevelt Hall and, along with ICAF, virtually the only occupant of Fort McNair. But by the 1970s new Pentagon training programs, strategic centers, and plain old bureaucratic sprawl had created a raft of entities scattered around Washington with no permanent or centralized place to house them. Fort McNair was the solution. But how to do that, given the long preeminence there of the War College? The answer: maintain the War College's overall independence but place it under a broader umbrella organization called a "university," of which NWC would be one "college." Hence, the creation in 1975 of the National Defense University and the housing there by now of the NWC, ICAF, the War Gaming Center, INSS, and other entities.

That remained the situation until the mid-1980s, at least, and what became known as the Goldwater-Nichols reforms. Goldwater-Nichols was a far-reaching piece of legislation that called for reform in a variety of military areas. Of special interest to us were two of these. First, with regard to PME education, Goldwater-Nichols called for reform of the training program at the National War College and, most particularly, for a new pay scale and the recruitment of a prestigious and high-powered civilian faculty, as Kennan's original formulation had intended. Second, Goldwater-Nichols called for true jointness and, for the first time, made it clear that high-level promotions within the military would be contingent on the officer candidates having served in joint positions. It was under the new Goldwater-Nichols rules and requirements that I was recruited to NWC in the early 1990s.[5]

Meantime, some of the other basic problems and contradictions of the College, present from the inception, remained unresolved. They would continue to plague NWC and still do today. Among them:

1. What kind of an institution is the War College? Is it a real college (small c) or a military installation?
2. Who runs the College? Is it the Joint Chiefs, the Army, OSD, or the National Defense University (NDU) administration?
3. Is the College to be considered at the apex of professional military education in the United States, or is it just one war college among many?
4. What is the role of civilian faculty at the War College? Are they mainly independent actors as in a university or are they subject to military authority and discipline?
5. What is the role of research in the College? Is it to be encouraged or is teaching to be deductive, from prescribed texts, "by the book"?
6. Closely related is Kennan's original idea of the College as a place for innovative strategic *thinking*. Would that be revived or are we in a situation of orthodoxy where both the curriculum and the conclusions are already determined?
7. Finally, there is the perennial question of whom the College is for. Is it primarily for the Joint Chiefs? Is it mainly for the students? Or is it for the faculty and their research interests?

None of these questions has been fully answered over the now-sixty-year life of the College. Perhaps they cannot be answered; perhaps the tensions implied in the above will always be with us. But if and until they are resolved, the College will continue to exist in this curious limbo: neither a full-fledged college or university nor just a military base; neither a place for research and new thinking nor quite just a training institution; neither a place for free-wheeling free speech nor quite a top-down, command system. Can these contradictions be resolved? We will have to see.

NOTES

1. A wonderfully interesting historical and architectural volume is Allan Greenberg, *George Washington, Architect* (London: Andreas Papadakis, Publishers, 1999).

2. By pure coincidence, we had earlier bought a nice house in bucolic Brookeville, which is the next crossroads north of the better-known town of Olney, Maryland. And, on one of the town's periodic tours of the area's historic houses, we had been in the house and bedroom where Madison stayed. He was awaiting reinforcements from the Army in Baltimore but—talk about a Third World country!—the Army got lost en route and never arrived. Meanwhile, James Monroe, who would be Madison's successor as president, had ridden out to Brookeville to tell the President that the country was falling apart in his absence and, as the British went back to their ships, that the threat had ended. By this time, back in Brookeville, however, local historians have discovered that, while James Madison fled to Brookeville, the equally famous Dolly Madison went across the Potomac River to McLean, leaving them to speculate that maybe the Madisons had a modern marriage as early as 1812. Dolly's presence there explains why there is Dolly Madison Boulevard in McLean and why Brookeville lays claim to having been the "Capitol of the United States for one night."

3. This period in the history is especially well-covered by the following papers, mainly unpublished, available at the NDU library, or through the NDU website: "History of the NDU" (www.ndu.edu/info/history.cfm); "National Defense University Fact Sheet" (www.ndu.edu); James Stanfield, "Establishment of the National War College" (1955); Stanfield, "The Mission and Function of the National War College" (1955); Thomas A. Terwilliger, "An Analysis of the National War College" (Ann Arbor, MI: UMI, 1993, original dissertation done at William and Mary University, 1992); John Yeager, "Congressional Influence on NDU"; Susan Lemke, "National Defense University Genesis and Evolution";and James Keagle, "A summary of Major Activities and Their Evolution, Academic Year 1946–1947 thru Academic Year 1989–1990" (NWC, n.d.). I understand that former faculty members Janet Breslin and James Keagle are writing a history of the War College, but they have not been willing to share their manuscript with me.

4. The Army also has its own college, the Army War College in Carlisle, Pennsylvania. The autonomy of and rivalries between the services mean the Army was often more interested in the AWC than the NWC.

5. For the recommended changes, see Committee on Government Affairs, U.S. Congress, "Management of the Department of Defense" (March 23, 1983); Committee on Armed Services, U.S. Senate, Staff Report, "Defense Organization, The Need for Change" (Washington, DC: Government Printing Office, 1985); and Committee on Armed Services, U.S. House of Representatives, Bill Nichols, *Department of Defense Reorganization Act of 1986* (Washington, DC: GPO, 1986).

Chapter 2

Command Structure and Organization

The mission of the National War College, and of the National Defense University of which it has been a part since the mid-70s, is to ensure excellence in professional military education *and research* in the essential elements of national security.[1] I emphasize the "and research" part of this mission because, over the years and decades since George Kennan—"Mr. X"—was at the War College in the late 1940s, the research purpose had been steadily downgraded, deemphasized, and discouraged to the point where it had all but disappeared. To serious scholars on the NWC faculty, the lack of attention to or encouragement of research is a serious impediment and one of the main factors keeping the College from recruiting high-level scholars. We will have much more to say on this subject later in the analysis.[2]

More specifically, the mission of NDU and NWC is to prepare selected commissioned officers and civilians from the Department of Defense, Department of State, and other agencies of government for command, management, and staff responsibilities in a multinational, intergovernmental, and joint national security setting. The NWC curriculum, as we see in detail in the next chapter, emphasizes national security policy formulation, military strategy and operations, and the global geostrategic context. The "intergovernmental" emphasis comes from the fact the War College faculty and student body come from other agencies besides the Defense Department. "Jointness" is emphasized because at the National War College, more than in any other DOD institution, students are expected to transcend their parochial service loyalties and emphasize their ability to work together "jointly." And, recall, NWC, though an Army base, is in a direct line position under the Joint Chiefs and the civilian Office of the Secretary of Defense (OSD). And finally, "multinational" is emphasized because right from the beginning the College determined to bring in foreign officers to the extent that now international affairs officers may constitute up to 15 percent of the student body.[3]

The National War College, as we saw in the previous chapter, was created in 1946–47 as part of the post-World War II reform that also led to the creation of the CIA, the Department of Defense, OSD, and the Joint Chiefs. The War College thus predates the organization of its supposed parent organization, the National Defense University, by some thirty years. NDU was only created in the mid-1970s as part of a further administrative reform of that time.

It is somewhat unusual for a "college" to predate the founding of a "university"; usually the university is founded first and then separate colleges grow and are organized within it. The reverse order reflects, as so many things do, the unique position of NWC/NDU within the Department of Defense. The creation of NDU made little sense on pedagogical grounds; rather, it was purely an administrative reshuffling. The *facts* are that too many different operations had come to be housed at Fort McNair—or elsewhere around Washington, D.C. They were all competing for space, personnel, and budgets. The solution, purely organizational, was to create an umbrella organization, NDU, to house and oversee all these diverse other entities. The reorganization was purely an administrative one; it had nothing to do with the teaching, training, or research missions of the War College.

But this reorganization, while it made sense administratively, gave rise to additional confusion and tension. Technically, the War College, in terms of the accompanying wire diagram, Figure 2.1, is under NDU.

And while the War College is headed by a two-star commandant, NDU's president is a three-star, thus making clear the chain of command. But in reality, (1) the War College was there first and predates NDU; (2) it thinks

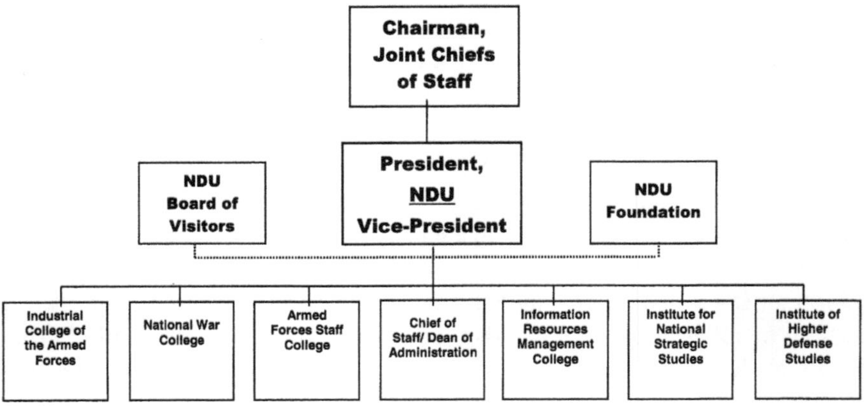

Figure 2.1.

of itself and often operates as an autonomous unit; (3) it has a very loyal and active alumni association, which NDU does not, and (4) it often has a direct pipeline into the Joint Chiefs that sometimes enables it to bypass NDU. I do not wish to dwell on this issue except to say that it provides at times some interesting tensions between NWC and NDU.

THE NATIONAL DEFENSE UNIVERSITY

During the period that I was there in the 1990s, NDU consisted of the following entities:

- The National War College, the most prestigious of the war colleges, with its long and storied history going back to the mid-1940s.
- The Industrial College of the Armed Forces (ICAF), with an even longer history going back to 1924. ICAF focuses on the training of senior military officers and civilian officials for the resource component of national power, with special emphasis on material acquisition, procurement, budgeting, and management, and the integration of these into the overall national security strategy. ICAF, housed in the Eisenhower building on the NDU campus, likes to think of itself as fully equal to the War College in prestige and importance, but in terms of students, faculty, and reputation, it is clear that NWC has the edge.
- The Armed Forces Staff College (AFSC) is a part of NDU and under the NDU "umbrella" and administration, but it is located on its own campus, the Norfolk Naval Station in Norfolk, Virginia, and, therefore, has considerable autonomy. As its name implies, it is a *staff college:* it is at a lower level than ICAF or NWC; it is for more junior officers; and it enjoys less prestige. There is some exchange of faculty and other personnel between the two schools but not much other interchange. The Armed Forces Staff College, as distinct from the staff colleges of the separate services, focuses on preparing its students for joint and combined staff duty and operations; that is also the logic—"jointness"—for its being housed under NDU. But there is another public administration principle that applies here, and that is that separation (geographic, by distance, three to four hours driving time) really means separation (in terms of functions, coordination, exchanges). For example, while I have lectured at ICAF on numerous occasions (right next door to NWC) and have good relations with many on its faculty, I have never lectured at AFSC.
- The Information Resources Management College (IRMC) is the capstone institution in the Department of Defense for information resources management education. Like ICAF and NWC, it offers courses in information resources

management at the graduate level. It prepares senior DOD officials for the joint (hence, its location at NDU) management of the information resources component of national security. It has four departments—Advanced Management, Information Management, Information Technology, and Executive Education—and is designed for senior DOD military officials (grades O-6 and O-5) and senior civilians (GM/GS-15 and -14). Interestingly, IRMC is headed, just the opposite of ICAF or NWC, by a civilian dean with a military deputy; its program and course of study, while, of course, integrated within DOD, are fittingly much more public administration-oriented than defense or strategic-oriented.

- The Institute of Higher Defense Studies (IHDS) administers a number of diverse educational programs carried out by NDU. These include the Capstone General and Flag Officer Course, the International Fellows Program, the Reserve Component National Security Course (RCNSC), and the NATO Staff Officer Orientation Course. I have lectured in the Capstone Course, had a number of the International Fellows (foreign officers) in my classroom; and have traveled around the country to military bases to lecture in the RCNSC courses (all of these discussed at more length in a later chapter). The only one of the components I have not had any connection with is the NATO course, primarily because I am typecast as a Latin Americanist and not a NATO specialist. IHDA is mainly an administrative unit to help manage special NDU programs; it has no students or faculty of its own and its staff is entirely military, headed by a full colonel or captain.
- The other non-teaching institute at NDU is the Institute for National Strategic Studies (INSS). INSS was created in 1984 specifically as a multidisciplinary research organization. It includes both military and civilian fellows and operates under both the NDU president and the Joint Chiefs. INSS is headed by a civilian academic with a defense/security background; its deputy is a military officer with the rank of colonel. INSS's research agenda (and, therefore, its staff—no tenure or permanence here) may vary depending on the shifting security/foreign policy agenda; it is staffed by approximately seventy-five specialists with expertise in various regional, country, and topical issues. INSS also administers the NDU Press, which publishes the Institute's research work in a variety of formats, and the War Gaming and Simulation Center, which for a long time was housed in an underground bunker until the mildew, water damage, mushrooms, and disease eventually overcame it and forced it to move. Of the various NDU colleges and institutes mentioned here, I had more contact with INSS and the War Gaming Center than any of the others.
- When I was at NDU, it had these four colleges and two institutes. Since then, in 2002, it has added the School for National Security Executive

Education (SNSEE)—already renamed as its fifth teaching arm, and by this time includes no fewer than fifteen other research program and training units. Some of these are rebaptized and reorganized from earlier units. Thus, the research centers, in addition to INSS, include the Center for Technology and National Security Policy and the Center for the Study of Weapons of Mass Destruction. Other programs include the renamed National Security Gaming Center; Interagency Transformation, Education and Analysis Program; and the Secretary of Defense Strategic Policy Forum for Members of Congress and Senior Executive Branch officials. Special components are the Capstone, Pinnacle, and Keystone programs for flag (general, admiral) officers, the Center for Strategic Communications, the rebaptized Joint Reserve Affairs Office, the NATO Staff Officer Orientation Course, International Student Management Office, National Security Education Program, and the Secretary of Defense Corporate Fellows Program. I do not even pretend to know what all these programs do or how viable they are. But I can report that sometimes NDU officials feel that, whenever the Pentagon comes up with a new program that doesn't fit anywhere else, it "dumps it" onto NDU to administer.

- Of special interest are some of the programs and centers aimed at training foreign security and defense officials. The one I know best is the Center for Hemispheric Defense Studies (CHDS), but there are now several others with a similar regional focus. CHDS was originally conceived to help educate Latin American armed forces officers in the principles of military subordination to civilian control and, by the same token, to help train Latin American civilians in the intricacies of running their countries' defense departments. Both of these are laudable goals and CHDS, therefore, as primarily an educational institution, was housed under the NDU umbrella—though its offices were located off-campus in the nearby Coast Guard Headquarters. But the Defense Department insisted, justifiably, on greater coordination of and control over the activities of these several regional training programs. At the same time, a recent NDU president, who has since left, saw these institutions as more operational than educational and placed them under the jurisdiction of the regional commands—in the case of CHDS, this meant Southcom. Thus, CHDS and the other regional centers not only lost their educational function but also the considerable autonomy they had enjoyed under NDU. But now that particular NDU president is gone, and CHDS has moved back onto the NDU campus in the newly completed Lincoln Hall. We will have to see if it remains under Southcom and, thus, retains its operational focus, or if some of NDU's educational and research activities now rub off on a revived CHDS.

NDU COMMAND STRUCTURE

NDU operates directly under the Joint Chiefs of Staff and its chairman. Since the Joint Chiefs are under the Secretary of Defense and the Office of the Secretary of Defense (OSD), one would expect NDU also to answer and be responsible to the Secretary. But it is rare for OSD or the Secretary to get involved in NDU activities; in my experience, only William Perry, Secretary of Defense under President Bill Clinton, has taken a more-than-passing interest in NDU affairs.

NDU is really more a creature of the Joint Chiefs than of the Secretary. That is because the Joint Chiefs, perhaps above all else, stand for "jointness," which is also the guiding principle on which NDU was founded. Both the Joint Chiefs and NDU have as among their principle purposes the integration and coordination of the previously separated services—hence, the emphasis on jointness. The Joint Chiefs, unlike the Secretary, have an ongoing interest in NDU; they select the NDU president; occasionally, the Chiefs will assign a new task, function, or even institution to NDU; and the NDU president reports back to the Chiefs. The assignment, sometimes arbitrarily or because there is no place else to locate them, of functions or even whole institutions to NDU by the Chiefs explains why NDU consists of such a diverse, even motley, collection of colleges and institutes.

The president of NDU is a three-star lieutenant general or vice admiral appointed by the Joint Chiefs on a rotating basis from one of the four military services. All of the colleges, schools, and institutes that are a part of NDU report to the president. It is a command structure.

Normally, the Chiefs look for a candidate for the NDU presidency who is not only a three-star but also who has some educational background in his career, either at one of the military academies, the staff schools, or the other war colleges. But that is not necessarily or always the case: the NDU presidency is a "billet" that needs to be filled and if there is a well-qualified candidate available at the right rank, so much the better. But it is still a billet that *must* be filled and if the only candidates available are not so well qualified, that is the way it goes. In the armed services, all the parts are supposed to be interchangeable.

In my experience, the NDU presidency has been occupied by two kinds of persons. The first are three-stars who are ready to be put out to pasture, are on the verge of retirement, and just want to finish out their careers on the lush green lawns and in the luxurious senior officer housing at NDU. They don't take much interest in running NDU, want to keep it on an even keel, and are passive administrators. From the point of view of the colleges and institutes housed at NDU, this is the best kind of administrator to have because he/she allows the colleges, which are long established and have institutionalized ways of doing things, to continue running their own affairs without outside interference or top-down decree-laws.

The other kind of president, the bane of NDU's colleges and institutes, is a heavy-handed one, full of fire and brimstone, who thinks he/she knows everything there is to know about military education or even education in general, wants to change and reform the entire institution, and especially seeks to put "those independent professors" in their place. From the colleges' and institutes' perspective, this is the worst kind of president to have. He/she will usually spend a year or two instituting grand, sweeping changes, fulminating at the scholars and teachers, and wondering why his reforms are not carried out. Meanwhile, the entities at NDU will go into a crisis mode trying to explain to the president why his plans won't work and, meanwhile, using every trick in the smart bureaucrat's book to frustrate these sweeping changes. I have lived under both kinds of presidents at NDU, male and female; believe me, the passive kind is *much* to be preferred over the activist one.

But the problem is even deeper than that. It has to do with *competence* and the ability to run a complex educational institution like NDU that must give at least some bow in the direction of being, and being governed like, a serious university. In my time at NDU, I served under three presidents: Vice Admiral John Baldwin (1989–92), Lieutenant General Paul Cirjan (Army, 1992–94), and Lieutenant General Ervin Rokke (Air Force, 1994–96). Of these three, only General Rokke, who had a Ph.D. from Harvard, served as a Professor of Political Science and Dean at the Air Force Academy, and was later President of Moravian University, had the experience and knowledge to serve as president of NDU. Moreover, interviews with former NWC deans and others closely associated with the University over the decades reveal that Rokke was the only qualified president in the entire thirty-five-year history of NDU.

The vice president of NDU is a civilian, a foreign-service officer of senior ambassadorial rank, appointed by the Department of State. This position is a bow in the direction of the mythology that NDU is not only "joint" but also "inter-agency." In fact, other than a close-by reserved parking space, the civilian NDU vice president is completely subordinate to the military president, has no or very limited authority, and really has very little to do. Though the accompanying wire diagram shows the vice president as directly under the president, and though NDU flyers indicate the president "consults with" the vice president, in reality the president is the sole authority at NDU and may consult with the vice president only if he wishes to. And most presidents, separated by that huge cultural chasm between military and civilian, DOD and the State Department, have chosen not to—except at ceremonial occasions.

Technically, the vice president is in charge of dealing with the small proportion of the NDU student body who are State Department foreign-service officers (FSOs). But these are mature men and women; they usually need little counseling. Nor does the NDU president normally, let alone regularly, consult with or seek advice from the State Department vice president. Occasionally, of course, there are disputes involving the FSOs and their relations to the College that the vice president must deal with, including negotiating with the University president. But these are usually administrative as distinct from policy matters; it is the Joint Chiefs who run NDU, not the State Department or its representatives.

Moreover, recognizing the unimportance of the post, State usually sends over for that position failed, or disgraced, over-the-hill ambassadors, former ambassadors whom it wants to put out to pasture, or ambassadors who need a few more years before they can retire at full pension. The post is so lacking in power and influence that I have seen the ambassadorial occupants of it literally roaming the NDU grounds looking for something to do. I have had ambassadors who filled this post, because they had so much spare time, drop in at my office to inquire about books they should read and for advice on what seminars they should sit in on. In considering the NDU vice presidency, one should consider what Vice President John Nance Garner once said about the U.S. vice presidency: "it's not worth a bucket of warm piss" (bowdlerized for many years as "spit").

The NDU Board of Visitors is an official advisory body whose functional and professional expertise spans the several colleges, institutes, and functions that make up the University. In my time at NDU, the Board had fourteen members, about half military and half civilian, the latter consisting mainly of businessmen and defense contractors. The Board meets once a year and provides advice and recommendations to the President of the University. A good deal of staff preparation goes into these annual Board meetings and surely it is good for the President to get outside and private-sector advice but, in fact, the basic direction of the University is set by the Joint Chiefs, and the Board is largely marginal to that role.

Another body worthy of mention is the NDU Foundation and Board of Directors. It was organized in 1982 as a not-for-profit, tax-exempt organization under paragraph 501(c)(3) of the Internal Revenue Code. Because the President and other officials of the University are prohibited by law from soliciting outside funds, the Foundation does that for them by soliciting grants and donations from the private sector, mainly from military veterans and NDU alumni groups. It seeks to provide funds to support the teaching and research functions of the University for which congressionally appropriated funding is not available. These may include endowed chairs, special seminars, conferences, fellowships, and other activities that go beyond the regularly budgeted activities of the University. It may be hard to believe that, with its $700 billion-dollar budget, the Defense Department also needs to raise private funds, but I have found that the NDU Foundation does useful work, not least in providing funds for my own professional and research travel. The Foundation has an Executive Committee consisting of five persons, a small staff at NDU, a group of nine or ten honorary directors (Alexander Haig, Henry Kissinger, Donald Rumsfeld, et al.), and from twenty to twenty-five directors, mostly ex-diplomats and military officers.

Below the President and Vice President in the administrative hierarchy at NDU is the Chief of Staff who doubles as the "Dean of Administration." I put Dean of Administration in quote marks because this position is not like a dean in a civilian university; instead, in military parlance, the Chief of Staff oversees all NDU administrative operations and functions as an XO (Executive Officer) to the President. The Chief of Staff runs the day-to-day operating functions of the University; he may also advise the President on management, policies, and procedures. The Chief of Staff functions on behalf of the President in tasking line organizations (this is Public Administration 101) to accomplish their responsibilities, coordinate activities, and implement policies and procedures. He also keeps the staff and the President informed of actions underway, reviews correspondence prepared for the President, and keeps the University running. A good Chief of Staff is essential to the smooth operations of the University.

The Chief of Staff when I was there was an Army colonel; he has a small staff that also includes the base chaplain and its protocol officer. In the 1990s the Chief of Staff oversaw thirteen directorates and offices, most of whose functions are self-explanatory. These include:

Legal Affairs Advisor who also doubles as the University Ethics Counselor. In my tenure at NDU I consulted the lawyer on book contracts, tenure issues, and outside lecturing and consulting fees.

Long-Range Construction Office.

Public Affairs Officer. At one time an NDU President sought to force the faculty to clear *all* their writings and lectures as well as statements to the media with the PAO, but the faculty managed to shoot that one down.

Plans and Programs Office—special events, joint meetings, conferences.

Personnel and Administration Directorate—basically a personnel office; sometimes (in my case) able to overrule college commandants—more on this below.

Resource Management Directorate—finances, management.

Information Management Directorate—printing, publications, visual aids.

Logistics Directorate—procurement, facilities management.
Health and Fitness Directorate—nutrition, stress management, injuries, physical fitness.
Security Office—all security support functions; all personnel at the University must have a security clearance.
University Library—an excellent collection of contemporary readings, rich special collections (Colin Powell, et al.), the papers of NDU administrators, a superb staff. But a facility that is seldom used by the students or faculty.
Academic Affairs Directorate—advises on jointness issues and University-wide educational programs, of which there are few.
Reserve Affairs Advisors—under Academic Affairs but important for supervising National Guard and Reserve Officers programs, in which I and other faculty participated.

This largely completes our brief survey of the administrative structure at NDU—at least as it was when I was there in the nineties. At various points later in the book we will have occasion to comment more on various of these levels and offices within NDU. But herewith are a few preliminary comments.

First, while the structure and organization have a certain superficial resemblance to a university, the reality is not that at all. That is, NDU is made up of several "colleges," "centers," and "institutes"; it has a president and a vice president; it has a "dean of administration"; and it engages in teaching and research. But in actual fact, NDU remains a military base; it operates on the basis of a top-down command structure; and the three-star in command governs on the basis of decree-laws or orders. The reality is that the academic departments at NDU have no real power; there is no faculty senate as in a civilian university (I once suggested this to my dean; the idea was shot down immediately); and subordinates, including faculty, are "tasked" (ordered) to carry out their mission. Let there be no mistake about it: this is a military command system and is not at all like a civilian university.

Second, NDU keeps changing. Whether it will move in the direction of becoming more like a real university is hard to say. A recent innovation is the creation of the office of a "provost," headed by a civilian, under the (still military) president. The new provost's job is to oversee the civilian and academic functions, including faculty appointments, at the university. In making this change, NDU is following the example of the Naval Post-Graduate School in Monterey, California, which has had a civilian provost for some twenty years. But the person chosen for the NDU provost's position, I am told, has no military background and has never worked under a military command structure. So this person will have to feel his way, and we will have to wait and see if any real power devolves upon this position, if it achieves some real authority independent of the command structure, and how it works in the long run.

The third point to make is that all of this discussion of the NDU command structure may be largely irrelevant. Recall that NDU was created as a kind of umbrella administrative structure presiding over a rather motley collection of unrelated functions and units; the main colleges (ICAF and NWC) long predated the creation of the University and had a proud history of their own; and that NWC—at least when I was there—continued to function quite independently from NDU and sometimes (as in the case of the PAO seeking to muzzle the faculty) in fairly open defiance of it or the president's directives. But inevitably when you create an umbrella organization like this, there is bound to be a certain amount of coordination between the component units as well as a centralization of functions. Remember also that, while NWC is headed by a two-star commandant, the NDU president is a three-star. When push comes to shove, the two-star has to obey the orders—if it comes to that—of a three-star. So gradually, over the last three decades—and the same thing has occurred in civilian universities—there has been a considerable centralization of functions, greater "coordination," and more standardization among the distinct units that make up NDU. And that also means a gradual enhancement in the power of the NDU president and staff, at the cost of a certain loss of power and independence for the previously autonomous NWC.

THE NATIONAL WAR COLLEGE

The structure of the National War College closely mirrors that of the umbrella organization, NDU. Or maybe we should say, since NWC was there first, that NDU mirrors NWC. In either case, the structure is top-down, military, authoritarian, a command system. See Figure 2.2, which presents a wire diagram of the College.

The Commandant of the War College is a military officer of two-star rank. The position is a nominative one within the military and it rotates among the Army, Navy, and Air Force. Normally, as with the NDU President, the Joint Chiefs would look for a two-star for this slot who has some prior experience teaching in or administering another Professional Military Education (PME) school. But that is not always the case; the billet has to be filled and, therefore, it's quite possible you'll get a Commandant with no experience in higher education and even one who is hostile to educational and War College activities. I have worked under both kinds.

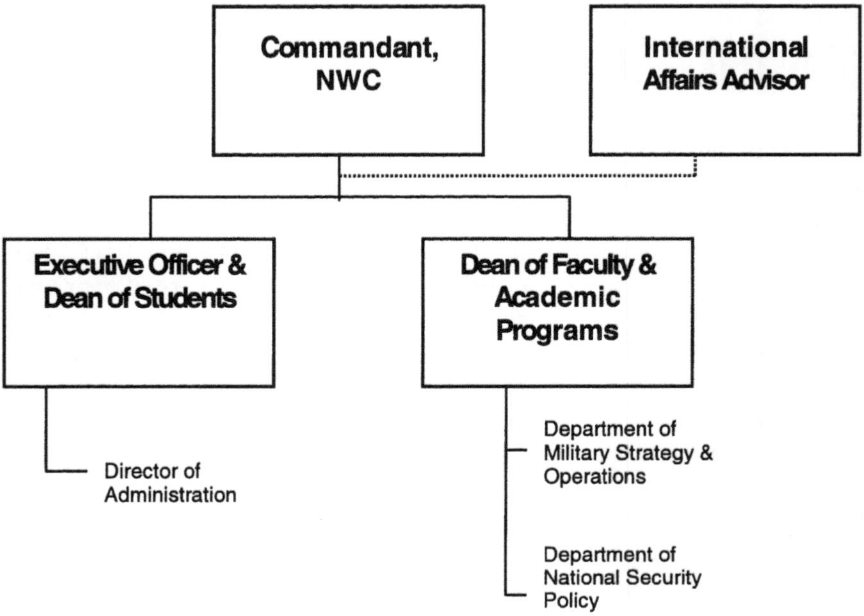

Figure 2.2.

As the name implies, the commandant is the "commander" of the NWC "post." The language is military; the structure and chain of command are military; and the system is top-down and authoritarian. The commandant is in "command" of all matters relating to the War College. That includes teaching, personnel, students, the curriculum—everything. They are all his responsibility. The commandant also has authority over appointments, though ordinarily those are handled at lower levels and the commandant merely ratifies decisions made below his rank. Ultimately, the NDU president has authority over the commandant and can overrule him, but rarely does that happen.

And just like at NDU, there are two kinds of commandants at NWC. The one type is a general or admiral who is content with his career, ready to retire, and is only looking for a quiet pasture where he can live comfortably, maybe play a little golf or tennis, and go quietly to live on his pension. The other type is feisty, wants to make his mark, and thinks he can remake NWC in his image. This second type is often arrogant, likes to order people around, and seldom has much use for the curriculum or the civilian faculty. Again, I have worked under both types.

I worked under three commandants at NWC: one Navy, one Air Force, and one Army (not necessarily in that order). The first was feisty, opinionated, and wanted to make his mark on the College. He was definitely of the second type discussed above, seeking to remake the College in his own image. He thought of himself as a tough, no-nonsense commandant who allowed no questioning and no dissent within "the ranks" (meaning us, the civilian faculty). I would say the other two commandants under whom I worked were a combination: tough and authoritarian when they wanted to be but open to at least listening to the civilian faculty. *None* of the commandants I worked under were of the peaceful, ready to put out to pasture type.

The first commandant I worked with (in military parlance you say "served under") was not only a top-down type but also had a temper. I have seen him struggle to keep his temper in check. Fortunately, when he exploded, he would explode at his Executive Officer (XO) and not, usually, at the faculty, and never at me. He also had strong ideas about teaching, the curriculum, and the value of the civilian faculty—most of them negative. He looked down on the civilian faculty, was dismissive of them, and was unalterably opposed to any permanence (tenure) for faculty members. He referred to the faculty at NWC as a "pass-through faculty," echoing the military conception that all the parts are interchangeable and, therefore, you don't need to keep a permanent academic faculty with real expertise on board. At the same time, he told me once that "all military officers are teachers"—a true statement, of course, when officers are dealing with the enlisted men under them, but sadly lacking in veracity if one considers the high level of specialized knowledge and expertise needed to teach effectively at the post-graduate National War College. If the parts are interchangeable, however, it follows that specialized knowledge and competence are not needed either. Hence, in his view, we did not need a permanent faculty, let alone tenure. Instead the faculty should be rotated quickly in and out of the institution.

Given these negative comments about the commandant, it may be surprising to learn that I used to play tennis weekly with him, along with some other NWC faculty. We played with him because, as Commandant, he had access in winter to the indoor courts at the nearby Navy Yard. For his part, even though he visibly disliked the faculty, the commandant needed tennis partners. He was as crusty, authoritarian, and arrogant on the tennis courts as he was in the halls of NWC. We all loved to beat him in tennis, and especially to smash the ball into his body at the net, since that was our only way to legitimately get even for his disdainful treatment of the civilian faculty at the College. And in the process of playing with the commandant, we sure learned a lot about him: all his prejudices (including against us), his arrogance, and his condescension. He was the wrong person to serve as commandant of the National War College.

The next Commandant I worked under was more easygoing, more relaxed in the position, more intellectual, and more self-confident. It was he who brought computers and the internet to NWC and insisted that we switch over to a paperless office. This commandant mingled with the faculty more easily than did the first one and was not so complexy in his relations with them. But he could also be tough: he was unalterably opposed to making concessions to the faculty and especially to the institution of tenure (a perennial issue at NWC) for the civilian faculty, arguing: (1) as Commandant, it was his authority to appoint, reappoint, or not reappoint; (2) changing security issues called for a changing or rotating faculty; and (3) the "parts," including instructors like us needed to be interchangeable. There is logic to these arguments, but it is a logic that applies more to a military base than to an academic institution.

I had less contact with the third Commandant under whom I served [4]at NWC. That is because I was gone, away on a leave of absence, at the time he arrived at the College, and I never did get to know him well. Indeed, that was part of the problem I had experienced in getting my leave of absence approved in the first place. For, military style, the outgoing Commandant did not want to agree to a contract for me that both approved my leave *and* reappointed me to a new contract for the coming year. Because that would "tie the hands" of the incoming Commandant and prevent him, while in command, from being the one responsible for making his own appointments. After all, that's what Commandants do: they're in *command!* In this again, one can see how the military mind works, how its logic differs from the logic of academia, and why on this, as on many other issues, it's so difficult to bridge the military and the academic cultures. Eventually, the issue was resolved: I got my leave and my contract was renewed in advance, but neither the outgoing nor the incoming commandant liked the procedure.

The fact I had less contact with this third-commandant-in-five-years did not prevent me from going to see him, just before I left the War College and went back to academia and my think tank life, to tell him what was wrong with the College and how it ought to be fixed. I'll save that full story for later in the book. Suffice it here to say that in engaging in this semi-rebellious act, I scrupulously followed "the book," going up the chain of command, venting at my department head first, then the dean, finally the commandant. By the time I got to the commandant, I had my delivery and punch lines down pat; it was a pretty strong performance. What surprised me is that, instead of arguing with me as I expected to happen, the commandant took copious notes, writing down everything I said. It was as if everything I said to him was new—and maybe it was, even though the issues had been around NWC for years. Perhaps he had never heard an unadulterated faculty viewpoint before. I came away from the meeting thinking that, in the military command system, maybe I was the only person who had ever spoken to him so frankly and critically.

And then—surprise of all surprises—quite a number of the recommendations I made then became official policy at the College. Sometimes, even in the huge and cumbersome bureaucracy that is the Defense Department, you *can* make a difference.

I've gone through the records and interviewed quite a number of longtime officials at the College about the NWC commandant position. In this collective memory, there has been only one good and qualified commandant in the last thirty years. That is Air Force General Perry Smith who had a doctorate from Columbia, had a significant teaching and publication record for a military officer, and had a sure and informed grasp of high-level education. Unfortunately for NWC, that experience has not been repeated; to a person, the commandants since Smith have been incompetent, uninformed on educational issues, poor administrators, and at best mediocre. None of the last seven or eight has had any previous educational experience; the services often select officers who have no educational experience or promise of advancement. The sad facts of life are that the services care little about whom they choose as NWC commandant and fill the billet with almost any old two-star who needs a slot. It's a shameful process and helps explain NWC's fallen stature. Moreover, since my time there the situation has only gotten worse, with the recent commandant, I'm told, having a fundamental distrust of the civilian faculty, with no intellectual curiosity, and presiding over an intolerable situation.

Parallel to the situation at the NDU level, the National War College has both a military Commandant *and* a State Department representative at the ambassadorial level who is the International Affairs Advisor. Our wire diagram shows this position as both *under* the Commandant *and* parallel to it. And that is probably an accurate depiction. The Commandant is clearly the person in charge at the War College, while the International Affairs Adviser functions more like a deputy responsible only for the 25 percent or so of the student body that are civilians, mainly State Department FSOs. And since these are, again, forty-year-old and mature persons who do not need a lot of advice, the International Affairs Adviser in my experience serves mainly as an adviser for those few foreign-service officers who need and want career or assignment advice. He may also advise the Commandant on the civilian and foreign-service side of the War College.

The International Affairs Adviser when I was there was dual-hatted. When I first went to the College, the International Affairs Adviser was also head of the Department of National Security Policy. This dual title and position went back to the early days of the College when Kennan, then with the State Department, was both the International Affairs Adviser *and* the head of the National Security Policy Department. It made sense then; it no longer makes sense now.

When I first went to NWC, the International Affairs Adviser and head of my department was a career ambassador. He was an able administrator, politically shrewd, and both sensible and sensitive in dealing with a crusty commandant. But he had limited academic background, had never taught a course, had no scholarly or research background, lacked a Ph.D., and, therefore, should not have been the head of a civilian academic department. As my department head, he was obliged to read, approve, and sign my course syllabus before I could teach the course. That is the mark of a military institution; in civilian universities, it would be very unusual for a department chair to sign off on your course syllabus before you could teach it. That is one part of the way(s) the command structure at NWC policed and controlled its civilian faculty. The department head, wisely, never gave any of the civilian teachers any trouble over their courses, but the procedures are so old-fashioned—plus he was not equipped to judge the academic content of the syllabus.

Just below the Commandant at the National War College, and with an office adjacent to his, is the College Executive Officer (XO) who doubles as the Dean of Students. The XO, normally a full colonel or of equivalent rank, makes the College run on an everyday basis and is an all-purpose trouble-shooter.[5] It is to him that everyone goes if there is a major foul-up; the XO must fix whatever problem arises. In my days at NWC, the XO was a no-nonsense Army colonel, Kirk Lewis, who had a BA from Washington State University, an MS from Kansas State, had graduated from the Command and General Staff College at Fort Leavenworth, and had spent a year at Harvard as a National Security Fellow. Kirk truly "made the trains run on time" but, with his academic background and as a counter-weight to the then-commandant, got along well with the faculty as well. Future XOs who lacked Kirk's combination of administrative skills, political sensibilities, and ability to stand up to the commandant were less successful in the position.

The XO also doubles as Dean of Students. That meant that, whenever a student had a gripe with the program or a faculty member, he/she would complain to the XO. The XO would then have to "fix the problem." The trouble is, in NWC's and DOD's order of priorities, the students, as future flag officers, are more important than the faculty who, recall, lack tenure and were referred to by the Commandant as a "pass-through" faculty. The XO would, in these disputes, then have to walk this delicate tightrope between his loyalty to the faculty and his obligations to the students, backed by the Commandant. It was not an easy position to be in. We shall have more to say in the following chapter about this curious inversion of priorities where students are more valued than faculty.

Alongside the XO and the Dean of Students in the NWC organization chart, and supposedly equal to him, is the Dean of Faculty and Academic

Programs. But I never thought the Dean of Faculty was equivalent to the XO and the Dean of Students. Part of it was the problem alluded to above: that on almost all issues the Command structure at NWC would side with the students over the faculty. Part of it was function: the XO, as an all-purpose problem-solver, simply did more and got into more things than did the Dean of Faculty. And, third, was location: the XO had his office right next to the Commandant and had direct access to him, whereas the Dean of Faculty was down the opposite corridor and was far away. It is an old Public Administration principle: location and access make a major difference.[6]

I served under three Deans of Faculty at NWC. The first, Roy Stafford, was undoubtedly the best. An Air Force colonel and with a Ph.D. from the Fletcher School of Tufts University. Roy had recruited me at NWC. He was a rare person at the College in that he understood intimately and from the inside both the military and the civilian university aspects of the College. More than that, Roy operated as a pluralist, one who could see several points of view, and as a conciliator and consensus-builder—as distinct from the chain-of-command structure that simply issued orders. Roy *loved* the College, staying on through his military career and even after his Air Force retirement as a civilian faculty member. All of us on the faculty felt confident that, as long as Roy was Dean, the College was in good hands and we would be protected. But from the Command's point of view, Roy was often seen as pleading the faculty's case over that of the students and even, at times, the commandant. After my time there, Roy was let go by the College in a set of circumstances that can only be described as disgraceful for NWC.

Later deans of faculty were not nearly so successful as Roy had been. Recognizing the sensitive nature of this position and that the dean of faculty had to relate well to the civilian faculty members, the College usually tried to recruit for that position a military officer, at the colonel level, who also had a Ph.D. But having a Ph.D., by itself, did not guarantee good relations with the faculty. It depended on which aspects of the officer's career, the military or the civilian-academic, had had the most impact on him. Some military officers can go through years of graduate school, earn a Ph.D., and yet never really absorb much of the academic culture. They remain, preeminently, military officers rather than academics. Some in fact even while earning a Ph.D. become hostile to the entire academic enterprise. We had several of these at NWC.

The other issue is whether the dean of faculty is willing to stand up to the commandant and defend his faculty in the face of a sometimes hostile command structure. Remember the commandant is a two-star general or admiral while the dean of faculty is only a "mere" colonel. So standing up to the commandant too forcefully or with the wrong tone of voice can be, for the dean of faculty, a career-ender. Somehow Roy, by his quiet but persistent arguments, was able to pull this off, while his successors were less successful. In fact, I got the feeling that some of our succeeding deans of faculty were so cowed by the commandants under whom they served that they put their own career considerations above that of the College. They completely buckled to the command authority.

Roy operated as a pluralist, allowing all voices to be heard and then making decisions that none of us doubted were motivated by considerations of the best interests of the college. That is usually the norm in good academic institutions but at the War College, with it chain-of-command system, such an attitude and way of operating is quite unusual. Other deans under whom I worked were less pluralist, less consultative, and more arbitrary, top-down, and authoritarian. More in keeping with the military character of the College. Though the title was dean of faculty, we perceived them as taking the command's side against their own faculty. Not serving the transmission-belt function, both top-down *and* bottom-up, as Roy had done. Or else, in the case of other deans of faculty, being so bland and neutral that they were unable to convey the faculty's concerns up to the command structure *or* explain the commandant's logic for doing or not doing certain things to the faculty. The dean of faculty position was an important, sensitive, and key position at the NWC; it requires a lot of skills and background to bring it off successfully. But after Stafford we were less successful in the deans appointed.

There are, under the dean, only two "departments" at NWC. Again, we see the parallels to an academic institution, in the form of the two departments, even while the realities are often quite different. The Department of Military Strategy and Operations had primary responsibility for teaching the "Foundations of Military Strategy" and "Military Strategy and Operations" parts of the Core Course, as well as some of the Advanced Seminars on military or strategic affairs. The Department consists entirely of military officers at the colonel level—except for the popular and irrepressible Ilana Kass who earned her Ph.D. at Hebrew University in Jerusalem and would regale the students with tales of the exploits of the Israeli army and intelligence services. The Department was headed by a full colonel, always a military man. No generals! The presumption was—a mistake—that flag officers were above learning anything new from the NWC course, *and* that generals were somehow above the teaching function. Only colonels or civilian faculty—"mere professors"—did that.

The fact that all the instructors in the Military Strategy Department were colonels or the equivalent means they were only one rank above their students—and sometimes of equivalent rank. Almost all these colonels had Master's degrees (that is now almost a requirement of making the rank of colonel) but only one or two had a Ph.D. Almost none of them had ever taught before, nor did they have the expertise to teach advanced or specialized courses at this high level. The military faculty lacked both academic and teaching experience. This was an ongoing dilemma for the War College: it was the separate military services that assigned people to teach there and sometimes, when Roy Stafford was dean and at his urging, we got officers with an educational background, but most times that was not the case. The system was haphazard and yet, because it was the services who selected the instructors, there was little NWC could do about it except sometimes lobby and negotiate with the services over who was selected to teach. The overall result, which got worse over time, was a military faculty that had almost no teaching background or academic qualifications. The system also reminds me of an earlier commandant's comment that "every military officer is a teacher," as if, as in a machine, all the parts were interchangeable.

The Department of National Security Policy (my department) was the second of the two War College departments. We were primarily responsible for the "Foundations of National Security," "National Security Policy Process," and "Geostrategic Content" parts of the Core Curriculum, as well as for Advanced Seminars. Roughly two-thirds of the Security Policy Department faculty had Ph.D.s; the others consisted of senior ambassadors, senior (SES-Senior Executive Service) employees from CIA and other agencies, and some military officers with the rank of colonel.

Most of the Ph.D.s at the War College were not true, university-based academics; rather, they were recruited mainly out of government service ranks (CIA, DOD, State, Congressional staff) and not from academia. That means that, while often quite brilliant, they did not usually have strong research, teaching, or publications records. Their degrees, like mine, were mostly from good public state universities, not usually from the most prestigious Ivy League institutions. And because they came out of careers in government service, they were used to working in large public bureaucracies, taking orders, and subordinating their private interests to the good of the institution. I believe that, at the time, I was the only person on the faculty to have been recruited out of an academic institution, although what made me attractive to them was not my academic background and publication record but mainly my Washington think tank and policy background. In general, it was believed, probably accurately, that university-based "academics" did not make a good "fit" at the War College.

It remained a mystery to some of us with an academic background why the College assigned military officers to our department even though they lacked any teaching, let alone research, background. Part of it was that the military didn't quite trust us civilians and wanted to keep an eye on us; part of it was the fact DOD didn't want to spend more money to recruit a larger civilian faculty; part of it was the notion, again, that every military officer is a teacher. But at the level of the National War College, a supposedly graduate-level program where real expertise is called for, you could not stick just anybody in the classroom with no specialized knowledge or experience. So we civilian faculty were asked to take some of these colonels into our classrooms and "mentor" them in how to teach, as if you can instruct persons who have never taught before how to teach in a couple of weeks of sitting in your classroom, and as if you can give them real expertise that has taken you years and even decades to build up. We tried our best to do this, but it is a badly flawed system and based on the flawed assumption that expertise is quickly learnable and all the cogs in the machine are interchangeable.

When I first went to the War College, in keeping with the "two-hatted" nature of the position, we had a senior ambassador from the State Department as our department head. He, as well as the rest of us in Security Policy, recognized that this was a bad fit, but the ambassador, as a good diplomat, made the best of it, signed off on all our course syllabi, and never gave the academic faculty a hard time. But even when I arrived, there was already a movement underway to change this system: to have the Ambassador continue to serve as the International Affairs Advisor to State Department students and as Liaison with the Commandant, but to have a different person appointed as head of the Security Policy Department. As that ambassador was about to leave, this was a good time to make the change.

Now the question was: Should the new department head be a military officer or a civilian? Most of the faculty in the Department was civilian so that, if there was anyplace in the War College where it was appropriate to appoint a civilian, this was it. The faculty, myself included, lobbied for one of our own, a civilian academic, to fill the post; but instead the College selected another colonel. He was a nice fellow but, lacking a Ph.D., was not up to the job of department head. Whenever I needed anything—travel funds, typing done, my course syllabi cosigned—he would grant or procure it without ever a complaint. But there was, without the academic credentials, little leadership for the Department. Why have a military officer as department head—except to keep an eye on the civilians—if he provides no leadership?

The next time the department chairmanship came up, we lobbied for a civilian head. Given the problems of the absence of research opportunities, they heavy teaching load, and the top-down and often authoritarian structure of the College, we felt a civilian department head would understand our concerns better than a military office did. And, since he was part of the power structure, that he would more sympathetically represent our views to the commandant and to the three-star president of NDU.

After considerable hemming and hawing as well as debate, we eventually got our wish. A civilian was appointed head of the Department of National Security. All of us on the civilian academic side cheered the appointment. But we were to be sorely disappointed. In office, the civilian head proved to be worse than any previous military head I'd had. He was, in our view, arbitrary, intrusive, and even more authoritarian than the colonels. We took to calling him, behind his back, "Napoleon" and worked out a system of warning signals when the head was coming down the hall to visit department members.

The new head and his behavior were the subject of great discussion among the rest of the faculty. He was one of us, so why did he behave this way? Had the office of head changed him or had he, seemingly a nice, competent colleague, always harbored resentments toward his civilian colleagues? Was he, a holdover from the earlier regime, angry at or jealous of the new, higher-paid, Title 10 faculty? Was he a secret Napoleon eager for power and the ability to boss the rest of us around? Was he trying to curry favor with the military brass by putting the civilian faculty in their place? Alternatively, had he been ordered by the brass to put an end to civilian lobbying for tenure, writing time, and a more academic climate at the College?

Whatever the ultimate answers to these questions, the faculty was not at all pleased with the directions the department now took. Our new chair seemed to take pleasure in browbeating the faculty, presumed to give us orders as if *he* were a general, sent "observers" (spies) into the classrooms to make sure we were teaching correctly; and at times he seemed to be looking over our shoulders as we worked at our computers. He singled out some of the faculty as "trouble-makers," "dropped in" at my office unannounced and uninvited to tell me what I was doing wrong, and sought to enforce the dictum never enforced before by military chairs that any research and writing we did had to be done on our own or "private time"; i.e., at night. He told the secretaries not to type any letters or research materials for me; there was even the rumor, which I had reason to believe but could not definitively prove, that he had tapped into and was reading our emails. Morale in the Department sank; I and others in the Department who could leave made plans to do so.

This was not a tolerable situation. We had wanted a civilian, now we had one and he was undoubtedly the worst department head in *any* institution, civilian or military, that I have ever worked under. When he walked the halls seemingly looking for someone to berate, we would try to get the advance word out that "Napoleon is coming." We all complained mightily to the dean (under whose instructions the head may have been operating) and eventually those efforts paid off: the department head was removed and quickly, replaced by another military officer. Our hoped for civilian head was now out and the colonels—for good or ill—back in. The moral of the story: if you want a civilian academic as your department head at the National War College, be careful what you wish for. Your wishes may come true—with dire consequences.

This concludes our consideration of the command structure and organization of the National War College. In the next chapter we discuss the War College program and curriculum, after which we go farther down the chain of command to discuss the student body and the faculty.

But before we do that, we offer a few concluding remarks about the command structure and organization:

1. The War College is really a *war* or *army college;* it bears only a distant relationship to a real college or university.
2. The system is top-down, authoritarian, hierarchical; lacking a faculty union or a faculty senate, there is no way voices low on the totem pole can make themselves heard. It is a military command system, definitely not a participatory democracy.
3. At every point in the system—Joint Chiefs, NDU President, NWC Commandant, XO, deans, department heads (except for that one brief, unhappy experience recounted above)—it is the military command that is in charge, not civilians.
4. Those aspects of the College that appear to be academic or resembling of a university—civilian vice president, civilian international affairs adviser, deans, academic departments, and the like—are only bows in the direction of civilian control and of an academic institution. None of these positions has any power as compared to the military command. Or if it does, the *appearance* of a university-like structure is used to obscure or even disguise that it is really a military institution.

NOTES

1. In writing this chapter, I have used the War College's own website, the "National Defense University's Fact Sheet," the official "History of the National Defense University" also from the website, and especially the revealing *National Defense University Catalogue 1992–1993* (Washington, DC: Fort Lesley J. McNair, 1993).

2. Given that research, writing, and publication are frowned upon and often actively discouraged at NWC, it is quite remarkable, and even heroic in my view, that some faculty members there are as productive as they are.

3. The first foreign officers were from Great Britain and the NATO allies. I was at the War College in the early-to-mid-1990s when the first Russian and East European students were admitted. One of my students was a Hungarian who presented me with a piece of the Iron Curtain. He later rose to become chief of staff of the Hungarian

Army. When I lived in Hungary in 2001 doing research for a monograph on EU and NATO enlargement, he was my best contact within the Hungarian military. He gave me access to the Defense Ministry and took me out on military training exercises. When I then visited the U.S. embassy, the military attaches were often amazed that I knew more than they did about the internal workings and problems of the Hungarian armed forces.

4. In the armed forces you say "under whom I served"; in a university or academic setting, you would say "with whom I worked."

5. Recently the XO and Dean of Students post was reorganized as a civilian position. However, hedging its bets, NWC appointed a retired Marine to what was to have been a civilian slot.

6. The recent reorganization of the XO position to make it a civilian post also meant a decline in the dean of faculty's status since he is now in reality subordinate to the XO/Chief of Staff. I deem this a very bad and regressive move on NWC's part.

Chapter 3

The National War College Program

The National War College is like a college—and it's not. On the one hand, it is located on a beautiful "campus" in the heart of Washington, D.C.; it has nicely manicured grounds; and it occupies a sumptuous red-brick building—but without ivy. It has all the trappings of a college—a library, classrooms, a president, provost, dean, and department heads. It has professors, "students," and a curriculum. It offers courses, assigns readings, and holds seminars. It is an instructional institution. At times, it has offered academic degrees, and I hold one of these—a Master of Science in National Security Policy. I list that degree and the teaching commendations I was awarded at NWC proudly on my c.v.

But at this point, the similarities with other colleges and universities end. For one thing, that bucolic "campus" where NWC is located is in actuality a military base. Second and more than that, it is an *Army* base, a theme the implications of which we will get into later. Third, beneath all those trappings of academia—provost, deans, and the like—NWC operates on the basis of a top-down, military command structure, not on the basis of a bottom-up academic institution. Fourth, the courses and seminars it offers tend to be based on established texts and deductive reasoning; NWC functions and teaches "by-the-book" in more ways than one, as if it were operating from a military training manual. It goes without saying that, unlike a good college, there are no courses in chemistry, physics, or English at NWC, although the latter would certainly be desirable to help with the students' writing abilities—better, the lack thereof. It is also the case that there is no tenure at NWC, only limited academic freedom, and, given the "by-the-book" orientation, no emphasis on basic research. Although NWC has at times offered its year-long program as Master's-level and degree-earning, there are no research papers or theses expected at the College. I would say that it is more of a training program not all that far distant from a technical school and less akin to a liberal arts college or university.

The National War College is, in fact, a mix, a hybrid. It is partly a college and partly a military training program. While it has no tenure, it has moved toward longer-term contracts for its faculty. It says it favors research and academic freedom but, as a military institution, it does not understand or quite "get" either of those two essential features of a university. It encourages give-and-take but only up to a point. As a dependency of the Joint Chiefs of Staff and a training program for the senior officer corps, it worries more about its students than it does its faculty, treating the latter as expendable and the former with kid gloves. The War College seeks to integrate itself more closely into the Washington community of academic institutions, but at heart it is part and parcel of the Army and the Defense Department.

As these comments imply, the National War College is chock full of tensions and contradictions. That is, in fact, true of all the PME (Professional Military Education) schools, ranging from the service academies (West Point, the Naval Academy, the Air Force Academy) to the staff and command schools for junior officers (such as Fort Leavenworth), to the war colleges (Army, Navy, Air Force, Marines), to the National War College. Each of these schools has somewhat different structures and procedures; each often reflects the personality and inclinations of its commanding officer—remember, these are military bases organized on a command basis. Each of them also manages the inherent tensions referred to above in somewhat different ways. But in all of them these tensions between a military institution, on the one hand, and a teaching and research institution, on the other, are always present, every day, in everything we do. I have myself concluded that these tensions will always be present because of the two very different cultures and traditions—military and civilian-academic—that come together in these institutions. The tensions can at times be mitigated, and there are good as well as bad ways to run these institutions, but the conflicts and contradictions will never be completely erased because they are built into the very nature of the enterprise that we are dealing with here: an institution that purports to be a "college" but also carries within the title the fact that it is part of the "war department," DOD. As one military colleague put it to me, perhaps indelicately, one time: "We teach how to wreck things and kill people more efficiently."

THE CURRICULUM

The National War College conducts what it calls a "senior level course of study" in national security strategy to prepare selected military officers and federal officials for high-level policy, command, and staff responsibilities.

The term "senior level course of study" is ambiguous. It can mean either that it's for senior-level military officers and civilians or that it's conducted at a level equivalent to a senior-level college course. The Defense Department prefers the first explanation, but I am not so sure. NWC directors consistently argue, for prestige purposes, that their program is at a Master's degree level and they have both sought and achieved accreditation at that level and, from time to time, offered a Master's degree in international security policy. However, my experience at the War College is that the courses, readings, requirements, grading, etc. are really at the level of a junior-senior undergraduate course in international relations.

Recall that the student body is three-quarters military, the majority being military academy graduates, most of the rest ROTC. Most of these guys as undergraduates were engineering, science, or technical students; in their careers they are tank drivers, infantry officers, boat captains, or airplane jockeys—most likely they have *never* had even a good undergraduate course in international relations, comparative politics, or foreign policy, nor have their careers in the military given them systematic training in these fields. They lack the background, the knowledge, and the theory to do advanced or graduate work in these fields. Hence, my assessment is that the NWC program is really at the level of a good undergraduate course. For the civilians in the program, only one-quarter of the class, the background is different: many of them were history or political science majors; quite a number of them already have advanced degrees; and they are better prepared for a higher-level graduate course.

But perhaps this is the wrong way to consider the program. Perhaps we should not—and this is my orientation—even try to compare the NWC curriculum to that of a civilian university. For this is the National *War* College, after all; its purpose and mission are to provide not just another IR curriculum but to train its students for strategy, security policy, and war. Therefore, it ought to be judged above all else by those criteria, not those of a civilian college curriculum. However, NWC itself prefers to think of itself as the equivalent of a civilian-level M.A. program and to compare and be compared at that level. So that is how we will proceed, and by those criteria NWC has a long way to go to measure up.[1]

The curriculum focus at the War College is on national security policy and strategy and emphasizes a joint and interagency perspective. According to NWC's own publications, the program is designed to expand and deepen the students' knowledge of national security matters and to sharpen their analytical skills. The academic curriculum consists of prescribed core courses, advanced studies, and student trips or what are called "regional studies." Teaching methods include lectures, seminar discussions, regional studies and issues, and special "student exercises" (including—one of the highlights of the year—an often wild weekend in Las Vegas!).

The most important part of the program is the core curriculum. The core, discussed in detail below, enables NWC to lay claim to offering graduate-level international relations training. It provides an understanding of the development and implementation of national security policy and strategy. It addresses both the domestic and international contexts within which policy and strategy are developed, examines the national security decision-making process, and focuses on the formulation and conduct of national security strategy, military strategy, and joint operation. Completion of the curriculum meets the educational requirements set forth upon the creation of the College in the 1940s for joint specialty officers.

The year-long (nine months) core program consists of seven segments.[2] Figure 3.1 shows the main subject areas and the time span of each segment of the Core Curriculum. Course I, Foundations of National Security, sets the broad framework and introduces the major concepts for the year's course of study. It addresses the nature of national security and presents a framework for analyzing U.S. national foreign and security policy. The classic issues in international relations and foreign policy are discussed here: the nature of power, idealism vs. realism in U.S. foreign policy, diplomacy vs. force as instruments of U.S. foreign policy, and so on. This is all familiar stuff to anyone who's ever taught IR or foreign policy, but not necessarily to military officials raised on a math/engineering curriculum.

I found this course, the readings assigned for it, and the level of instruction to be roughly equivalent to an advanced (junior-senior level) undergraduate course that I might teach on foreign policy. Moreover, even though the course is billed as discussing non-traditional security issues such as counter-

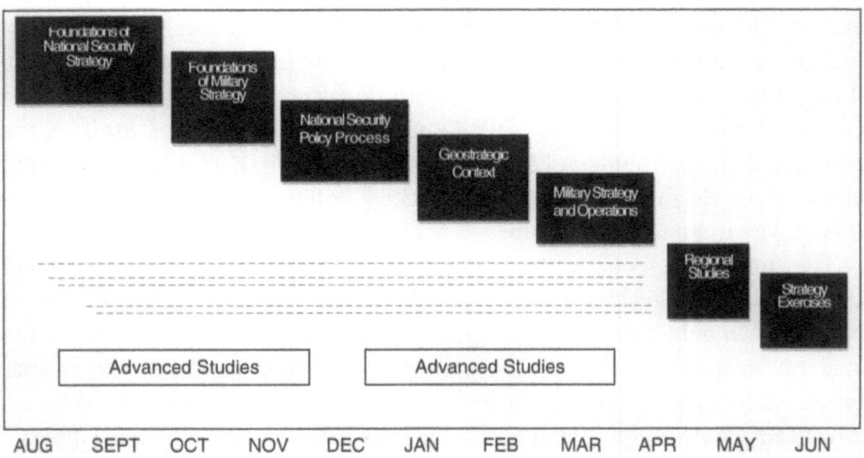

Figure 3.1.

insurgency, a-symmetrical war and nation-building, I found next to nothing in the curriculum on these crucial issues. In fact, the Department of Defense learned little from the Vietnam experience and, as we now know, was completely unprepared for the war and insurgency in Iraq.[3] For a course and program that purports to be at the graduate level, it should be noted that the entire year-long program has virtually no theory, an oversight that these practically minded students applaud but that is absolutely necessary if you are claiming to be a Master's-level course.

The course is divided into two segments. The first part examines the basic issues and concepts outlined above, the debate on U.S. foreign policy in the post-Cold War era, the choice of instruments (diplomacy, covert action, force) available to U.S. policy-makers, and the always-present trade-off, completely ignored in Iraq, between strategic objectives and available resources. Readers will be interested in knowing that, on the debate between realism and idealism in foreign policy, NWC students, not unexpectedly, come down strongly on the side of realism, but at the end of the course they are willing to concede—essentially a Ronald Reagan conclusion—that the support of democracy and human rights can, as in the case of the former Soviet Union, be used to achieve realist goals. On the instruments of foreign policy, the students come to recognize that they have a variety of "arrows" in their "quiver": diplomacy, propaganda (public affairs), covert action, military force. One of the enduring values of the course, with one-fourth of the student body made up of civilian agency representatives, is that the military officers come to recognize the usefulness of diplomacy and, as Kennan put it, "measures short of war," while the diplomats come to a better appreciation of how you use military force or the threat thereof to back up your diplomacy.

The second half of the course studies the key principles and concepts noted above through a review of national security strategies followed by American administrations since World War II. Essentially, during this part of the course we follow the outline and book of John Gaddis.[4] I found this part of the course useful from a teaching and learning point of view but, again, nothing new that would not be taught in a junior-senior-level course on national security. We began with Kennan's containment strategy and carried it through all the way to the end of the Cold War—but not beyond. What the students found particularly useful were not only the distinct iterations of the containment strategy but also the necessity (again ignored in Iraq) of doing a cost-benefit analysis of the gains to be achieved by certain policies versus the costs, monetary and otherwise, that would have to be paid.

Because of the rising importance of international economic policy during the post-Cold War, Clinton-era, 1990s, the War College had instituted a separate bloc run concurrently during Course I on economic principles as they relate to national security matters. The economic bloc is short, limited to six discussions, and, given the weak background in economics of the officers, about at the level of Econ 101. This short-course focuses on domestic fiscal and monetary policy, international trade and finance, and economic change in the former Soviet bloc. I did not teach in this bloc, but I did read the syllabus as well as the readings for the course.

Course II, Foundations of Military Thought and Strategy, is designed, according to the equivalent of the NWC "course catalog," to enable NWC graduates to understand the effectiveness of military power as an instrument of policy and the options available for the strategic employment of force in support of national policy objectives. With this knowledge, the argument is, NWC graduates will be better prepared to develop new strategic military plans or to critique existing plans. None of the civilian faculty, unless they had a military background, taught in this part of the course; but I endeared myself to my military superiors—because I was interested but did not know the literature well—by voluntarily sitting in on this course and doing the reading.

The course focuses on the tradition of military thought and strategy, from Sun Szu ("Keep your friends close; keep your enemies even closer!"), to Carl von Clausewitz (the military students are especially attracted to his notion, real in their experiences, of "the fog of war"), to "the present." But again, almost no attention is given to the ideas of contemporary revolutionary war: Lenin, Mao, Guevara, Fanon, General Gap. And certainly not, in the 1990s context, to asymmetrical war and Islamic fundamentalism.

The great theorists of war are presented from the perspective that they constitute a grand tradition of evolving military thought and strategy. The Course gives only a weak smattering of coverage to the catalysts (industrialization, globalization) that trigger new military thought or to the changing social conditions in which wars are fought. It also claims to emphasize the fundamental principles that remain constant regardless of changing conditions, but I did not find that was ever systematically or realistically done. Finally, the course allows the students to critique military tactics and strategy (Napoleon, Hitler) based on their knowledge of military theory; but, even though the students enjoy this exercise, it does not prepare them to deal with Vietnam-, Iraq-, or Afghanistan-style and contemporary a-symmetrical wars.[5]

Course III, The National Security Policy Process, deals with the domestic bases of U.S. foreign policy. The course addresses both the key organizations and the processes of national security policy decision-making. It is designed to help students understand the complexity of the policy process and how they can influence the outcome of that process. Two main themes dominate the course: what scholars call the "bureaucratic politics" of decision-making (competition, rivalries, overlapping responsibilities among foreign policy agencies), and the tension between the legislative and executive branches.

As could be expected of military men, the students at NWC like neither the bureaucratic politics model nor the political one implied in the focus on executive-legislative relations. They much prefer the "rational actor model"—clear lines of authority and responsibility; the president decides on the basis of a careful weighing of the options and a thorough cost-benefit analysis; the Congress generally goes along; and the bureaucracy (state, DOD, etc.) carries out the president's orders. Unfortunately for the orderly military mind, the process does not work that way. Politics, logrolling, political deals, paybacks, bureaucratic turf wars all get in the way of a rational, orderly process. Moreover, the course is directed and mainly taught by persons whose prior experience is, in the legislative branch, adding another bias to the training. It is important to know the congressional view and process but not to the exclusion of other main actors.

When the course finally turns to the executive side of the equation, there is less experience and knowledge on the part of the faculty. Few among the faculty have worked in the White House, the National Security Council, State Department, CIA, or other agencies. Furthermore, this is where the students do have expertise: most of them *have* worked in these agencies or DOD for upwards of twenty years. You do not need to teach them about bureaucratic politics (although the term may be new to them); they already know the bureaucratic system intimately from operating in it for all those years. Here we have a situation in which the students often know far more about the subject matter than do the instructors. My solution to this problem was to give the students some added general and theoretical readings on the bureaucratic politics issue (most often from my own book![6]) and then let them puzzle out how these apply to their own work and agency. The students didn't like the theory readings particularly, but I thought it was good for them anyway.

Course III concludes with a couple of sessions on the domestic political environment: changing American demography, political culture, public opinion, interest groups, budget considerations. Having written a whole book on the domestic basis of American foreign policy (which I argue is 90 percent of the policy—also disturbing to my military students), I believe this part of the course receives far too little attention. For increasingly in the post-Cold War ("no more enemies") Clinton Administration, foreign policy *was* domestic politics. The students got woefully short-changed on these issues—even though the *politics* of foreign policy-making is a theme they'd rather not think about.

Course IV, The Geostrategic Context, is in a sense "my" course. Essentially, it is a global comparative politics course—my academic specialty. The course, which starts after the Christmas-New Year break—what we in academia would call the "second semester"—examines the worldwide context in which U.S. policy is carried out. The course reviews the interests and perspectives of the major international players, the allies, the several developing areas, and potential enemies and problem areas. It is a global *tour d'force,* all done in about six weeks.

Usually the War College likes to get the Secretary of State, the Secretary of Defense, or at least an undersecretary to deliver the introductory Grand Overview for this course. But when on occasion those high officials didn't show up or canceled out at the last minute, I was asked to do the large *tour d'horizon* called for. This is my *forte:* grand, macro, comparative politics. A global view. Even though I knew the military command was skeptical of my being able to deliver a lecture at this high level, in fact I was able to bring it off, earning high marks from the students. In the Cold War's aftermath, I provided the big picture, talking about shifting geostrategic priorities, the decline of Russia, the rising importance of Japan, China, India, the Middle East, and Latin America, and the implications of all this for U.S. policy. Given that the first time I did this, I had only—after our scheduled speaker canceled out—a half-hour to prepare, I thought, if I do say so myself, it was a pretty good performance.

The introductory lecture, on geopolitics and the shifting balance of power, sets the framework for successive blocs on Russia, Eastern Europe, Western Europe, Asia, the Middle East, Latin America, and Africa. Of course, I lectured in the Latin America bloc as well, but that is familiar territory for me; more important in my view was the overview lecture in which I could set forth my ideas on the shifting priorities of global politics (Russia and Europe, down; Asia, the Middle East, and Latin America, up). The course was also supposed to examine the topic of insurgency in the Third World, but that always seemed to get short shrift and was not a high priority. Nor was there any attention during that period of the early-to-mid-90s to nation-building; in fact, with the unhappy and even disastrous Clinton experiences of nation-building in Bosnia, Haiti, and Somalia, NWC students wanted no part of that policy. They argued, with implications for later, often equally unsuccessful nation-building efforts in Iraq and Afghanistan, that those were essentially police or civilian functions, that the military was not trained or equipped to do nation-building, and that it was not their responsibility.

Course IV has one additional purpose: it is designed to prepare students for their May trips abroad, what the College calls "Overseas Regional Studies." In these trips the students have a chance to see first-hand what we have spent the year teaching them in the Core Course and Advanced Seminars: how other countries operate, their security and foreign policy perspectives, and U.S. interests in them. More on this below.

Course V is entitled Military Strategy and Operations. This course studies the development and implementation of U.S. defense policy and strategy. It mainly treats military operations at the level of the regional theater commander-in-chief and his/her principal subordinate commanders. The students seek to confront the challenges and issues involved in developing a military strategy and forces for an uncertain future. Because neither I nor others on the civilian faculty have any expertise on these issues, we did not usually teach during Course V. And unlike the earlier Foundations of Military Strategy course, where we also did not teach, in this course I did not sit in on the lectures or do the reading. I'm sure I would have learned a lot, but by this time of the year I was exhausted from my earlier teaching and wanted to get in some needed writing time. I'm sure there's an NWC regulation somewhere (see below) prohibiting that, too!

During Course V the students discuss the continued relevance of deterrence doctrine in nuclear and conventional conflict situations. They mostly study conventional war—what they know best—including its planning, execution, and termination. They consider the political and military impact of WMD (weapons of mass destruction). Finally, and in too abbreviated a time frame, they look at low-intensity warfare. But because the NWC students generally lack language, cultural, political, and nation-building skills and knowledge, they feel more comfortable with conventional warfare than with guerrilla insurgency.

The Military Strategy and Operations segment (Course V) ends the Core Course for the year. That is followed in May by two weeks of what are termed "Regional Studies." In Regional Studies, all of us, students and faculty alike, fly off to distant places, mostly but not always reflecting the Advanced Studies courses the students were taking. Between fifteen and twenty of these trips may take place, mostly with ten to fifteen students each with one faculty director and one assistant director. All expenses are paid.

The trips are usually scheduled for about twelve days, leaving on a Sunday and returning more than a week later on Friday. Since I taught the Latin America course, I was expected to lead the trips to Central and South America; other trips went to Africa, Asia, Russia, the Middle East, Australia, and Europe. As one might expect, France was the most popular destination, leaving open the question of whether there are really learning experiences or foreign junkets. Oh, yes, the base commandant gets to choose which trip he wants to accompany; usually the students try, if at all possible, to avoid that group.

At first, I was reluctant to sign up to lead one of these trips. The thought, as I indelicately put it at the time, of "playing nursemaid" for two weeks to a group of officers who didn't know the language, didn't know the culture, and had never been to Latin America, did not appeal to me. In addition, when snags developed in the planning and scheduling, the War College had the strange (to me) idea that I should use my own academic, personal, and political contacts to straighten out the problems. To the War College, that was part of "carrying out the mission," which had to be completed at all costs; to me, it was an imposition and a violation of professional ethics to use my own personal contacts, built up over years and decades of work and research in the area, for official, U.S. government purposes. This was one of many instances treated in detail below where the military and civilian-academic cultures would clash.

The students do most of the planning and scheduling for these trips; our job as faculty members is to serve as guides and instructors, once we get there, about the country or region visited. But, to illustrate the problem mentioned above, the first trip I took with a War College group was to Mexico and Central America. The students arrange with U.S. military attachés about such things as ground transportation, lodging, meals, and, hopefully, a challenging itinerary full of high-level meetings, excursions, and briefings. However, the head of the U.S. military mission in Mexico, a brigadier general, reflecting Mexico's importance in U.S. foreign policy, was giving our trip organizers a hard time. It was plain that, because of Mexican sensitivities, he did not want our delegation to travel to Mexico at all or, if we came, to do nothing but climb the pyramids outside Mexico City. Well, this will not do at all; we want a *substantive* itinerary. And it happened that the U.S. ambassador to Mexico at the time, John Negroponte, was a good friend of mine from the Central America political (as well as military) "wars" of the 1980s. So I placed a call to John, thus violating my own strictures just this once about using private contacts for government purposes, and suddenly everything fell into place. Not only did we quickly have a substantive and challenging program, but that prickly brigadier was chastised and my own reputation quickly soared in the eyes of my students.

I have written about these trips in detail elsewhere,[7] and need not repeat all that information here. Suffice it to say that on one trip we went to Mexico and Central America (El Salvador, Guatemala, Nicaragua, Costa Rica, and Panama), on another to South America (Argentina, Brazil, and Venezuela). The war College also generously sponsored trips by me to Russia, Asia, and Western Europe. *All* of these excursions proved to be extremely valuable and worthwhile. For, traveling with the U.S. military and interacting with the foreign militaries of these countries, you get to see and experience things that, as an independent civilian academic, you would never get to experience. Second, these military officers proved to be wonderful traveling companions: mature, fun-loving when that was appropriate, and serious and inquisitive when we were in high-level meetings. And third, because both transport, meetings, etc. were all arranged for us by the U.S. military attachés in the countries visited, we could really accomplish a lot in a short period of time. I saw and experienced so much on these trips that I consider them among the most informative of my life.

Returning from these trips, the students are required to write a briefing paper and report on what they learned. Their papers are then read and evaluated by their instructors and become part of a permanent record at NWC. For my part, I kept, as usual, a journal on these travels; these volumes to my mind constitute some of the most interesting of my journals.

The seventh and final part of the NWC academic year, accomplished in the last two weeks of May, is called "Strategy Exercises." I found this to be the weakest part of the entire program. The exercises are supposed to emphasize "active learning" and provide an opportunity to put theory into practice. These are three such exercises:

1. National Security Strategy Exercise. This exercise projects possible future environments of U.S. national security interests and develops objectives and a national security strategy for five-to-ten years ahead. This is the exercise that I worked on.
2. Military Strategy and Resources Exercise. This exercise develops a military strategy based on projected objectives and the threat environment and prepares U.S. forces to fulfill the strategy given finite levels of resources.
3. Crisis Decision Exercise. This exercise deals with the political, economic, and military dimensions of a crisis scenario set several years into the future. The exercise forces the students to access the nature of the crisis, determine U.S. interests and objectives, evaluate threats, and apply strategies and forces previously devised to secure those objectives.

The problems with this exercise are several. First, the two-week time period is too short a time to develop more than a very superficial report. Second, the fact these reports are done by groups of fifteen to twenty students compounds the first problem and gives rise to numerous problems of administration, coordination, consensus-building within the group, and difficult group dynamics. Third, the instructors (us), knowing how superficial these reports are, must nevertheless put our best feet forward and treat them as more serious than they really are. Fourth, the students themselves, at the end of this strenuous year, are eager to get this last hurdle over with and move on to their new assignments, giving often only limited attention and time to preparing their report. And fifth, these reports are finally evaluated by panels of retired officers, many of whom are crusty, are not up-to-date on what the War College is doing, and have had no previous contact with the students or faculty, leading to considerable miscommunications. All in all, these strategy exercises do not represent the best part of the War College's curriculum.

In addition to the five-part Core Course, Regional Studies, and the Strategy Exercises, students at NWC during the year take a minimum of four advanced courses, two in the fall and two in the spring. These courses are electives, which means only a small percentage of the student body takes any one of them. Most of these advanced courses have a regional- or country-specific content—China, Russia, South Asia, Latin America—or they focus on a specific topic: arms control, human rights, democracy, and so on. The courses meet in the afternoon (as opposed to mornings for the core courses) for an hour-and-a-half each. The courses enable the students to acquire more specialized knowledge on particular areas and topics—although my experience was that the students take courses on topics they have no experience in and know *nothing* about, rather than acquiring advanced-level knowledge.

To fulfill the four-seminar requirement, students may, if they wish, also take courses at NWC's neighboring institution, the Industrial College of the Armed Forces (ICAF) or, if offered, through the National Defense University. A great variety of courses—we can almost call them "electives"—are offered in this way. And the courses may vary from year to year depending on the interest and expertise of the faculty available. These offerings are evaluated in more detail below; suffice it here to say that the so-called "advanced seminars" are usually not so advanced at all.

Two other components of the War College program must be discussed here, though they are not part of the regular, year-long curriculum. The first of these is what is called the Capstone Course. The Capstone Course is a two-week course for flag officers (generals and admirals) who are brought to the War College for what is in essence a refresher course on aspects of national security that may have changed since they were students here or at another of the war colleges. I have lectured to the group on Latin America, but since Latin America is not viewed as a high-priority, national security area, those have not always been regularly scheduled.

The wives (should we now say "spouses"?) of these generals and admirals also receive a separate security briefing, and I have been asked ("tasked," which de facto means ordered) to speak to this group as well. So I went through my usual, fifty-minute spiel on Latin America and then asked if there were any questions. Silence filled the room, reflecting the fact the wives had no background or interest in the area. Finally, one lady raised her hand and said, "Mr. Wiarda, I see on your c.v. that you are a tennis player. Could you tell us what your rating is?" Nothing on Latin America or the substance of my talk, only tennis! Afterwards I dropped in to see the dean, Col. Roy Stafford, who also has a Ph.D. and is one of the all-around decent persons at the College, and said, "Don't ever ask me to speak to the wives' group again!"

The other component worth mentioning is the RCNSC—Reserve Component National Security Course. As the name implies, these are refresher courses in national security offered to reserve officers during their yearly training exercises. The faculty at NWC was asked ("tasked") to provide the personnel for these courses, which are held at military bases all over the country. I would lecture on Latin America; faculty colleagues lectured on other areas and issues.

At first, I resented this extra teaching obligation since it took me away from my research and writing time and agenda. But when you work for the Army, unlike in academia, you don't have a "research and writing agenda." You work *full-time* to accomplish *the mission,* which is to educate military officers on national security. In the military, your time is not your own; it *all* belongs to DOD. Coming from academia and the think tank world, this took some time for me to get used to; on the other hand, my military colleagues could not at all understand the notion of "free time" to do research and writing as practiced in civilian universities.

This also may sound dumb and egotistical but again, coming from the remunerative think tank world (AEI, CSIS), when I do "extra-curricular" lectures like this, I expect to be paid for them. I even made some crack to that effect when the dean of faculty (Stafford) at the War College asked me to do the RCNSC lectures. Of course, that notion was heartily guffawed right out the door. When you're *tasked* to do a lecture like this by NWC, you *do it*—no complaints, no questions asked, no honorarium. It's all part of your obligations as part of your job. That's an order, mister! Do it! Welcome to the military.

But once I started doing these lectures, I thoroughly enjoyed it. First, the men and women that you work with are very nice people—pleasant, educated, professional, competent, genuinely interested in the subject. The reserve officers I dealt with, however, often small-town fire chiefs in their civilian lives, police chiefs, businessmen, professionals, civil servants, tended to be a notch below my NWC students intellectually and, often, a little chubbier and out of shape. Second, I sometimes got to fly to these assignments out of Andrews Air Force Base in DOD's fleet of small executive jets, and I now understand why GM and other CEOs are loathe to give up their private jets. You go to the airport; the plane is revved up and waiting for you; there are no lines or security checks; the service on board is impeccable; and upon landing, a car waits for you right on the tarmac. Nice! And third, through this program I got to see military bases and parts of the country I never otherwise would have visited: the Pensacola Naval Air Station in Florida, Van Den Berg Air Force

Base in California, the Air War College in Montgomery, Alabama; Fort Benning, Georgia; the Miramer Naval Air Station, San Diego, California; Wright Patterson Air Force Base in Dayton, Ohio; and numerous others.

These visits completely transformed my notion of a military base, which over the years of the Cold War had become like elegant country clubs. I think the last time I had visited a military base had been after World War II when my uncles were discharged from the Army and my parents and I went to pick them up. My memories from then are of Spartan living quarters, Quonset huts, temporary barracks that looked like fire traps, unmowed fields, no frills, no luxuries. But today's bases are not like that at all. Particularly for the officer corps, these are plush places to live. They have beautiful, often waterfront, separate houses for the senior officers; golf courses that are professionally designed, tennis courts, outdoor and indoor swimming pools, luxurious officers' clubs with chef-made and subsidized meals, manicured lawns, bowling alleys, movie theaters, PXs with subsidized food and everything else; sometimes there is an entire commercial shopping center on the base. The bases have their own police and fire departments; crime is all but nonexistent; there are no property taxes to pay; the woes of the outside world are shut out; and the base exists completely autonomously from the outside community. They're more like a self-sufficient small city (but without the problems) than an old-fashioned (from my distant memories) military base. When I had the opportunity to visit U.S. military bases abroad in Spain, Portugal, Italy, and Germany, I found the same thing: quite luxurious living accommodations, complete self-sufficiency, and—disturbing to me, but that is another matter—very little assimilation by base personnel into the life, language, and culture of the local community.

I know that in this era of an all-volunteer army, we need to make military life attractive in order to garner new recruits. But country-club living conditions seem to me a bit too much. I wonder if those congressmen who vote such lavish budgets for the armed forces have ever seen one of these "country clubs" up close and from the inside.

WHAT WE DO EVERY DAY

A typical day for me while teaching at the War College began at 5:15 a.m. Washington traffic is horrendous; there are no expressways from our house in northern Montgomery County into Washington; and to beat the traffic we would leave the house before 6:00 a.m. To get to wife Iêda's job at the Library of Congress, we would speed down New Hampshire Avenue for twenty-three miles until it merged just inside the District of Columbia into

North Capitol Street. By actual count, there were seventy-seven traffic lights between our house and the Library; sometimes, if the lights were coordinated and the traffic not too heavy, which was rare, I could make seventy-two to seventy-four of the seventy-seven lights and arrive at the Library in forty minutes. Once there, it was another ten to fifteen minutes to the War College. I usually arrived around 7:00, a quiet time when I could get my teaching materials, research materials, and writing time organized for the day.

The students usually arrived between 8:00 and 8:15, going immediately to their office gathering rooms on the first floor (basement) of Roosevelt Hall. There they could catch up on the news from colleagues, share gossip, and complete (hopefully) the reading assignments for the day's discussion. Actually, I found that most of the students completed their daily reading: if they didn't, it showed and they looked bad in front of their colleagues; also, these are older guys and professional and, unlike my younger undergraduate students in civilian universities, their careers are on the line. Hence, you do the work assigned.

At 8:30 the bells ring and everyone, faculty and students alike, is required to gather in Arnold Auditorium. There are no unexplained absences here; it is your duty, obligation, and part of the mission to be present. If you are late for any reason, you cannot enter the auditorium; you must stay outside in another room and listen to the lecture over the intercom.

The 8:30 lectures are at the heart of the core curriculum. Often the lecturers (but not always their lectures!) are very high level: the vice president of the United States, the Secretary of State, Secretary of Defense, national security adviser, ambassadors from other countries accredited to the United States, undersecretaries and assistant secretaries of State and Defense (where real policy gets made), luminaries like Henry Kissinger, Zbigniew Brzezinski, Colin Powell, Brent Scowcroft. One of the highlights of the year was when Prince Bandar, the Saudi Arabian ambassador to the U.S. during this time, came to speak. Bandar always got a big round of applause because he was *macho,* smooth, accomplished, rich, powerful, a pilot, a "top gun" just like many of our colonels aspired to be.

As a teacher and lecturer, I have always felt it embarrassing to speak before a well-informed audience like this without having done thorough preparation. I've also learned how to "wing it," if necessary, if I don't have the time to prepare fully, but I much prefer (it's the Calvinist in me) the other method. Not so, apparently, with some of our speakers. Many of our speakers come fully prepared and are extremely informative: a well-thought-out lecture by a Kissinger or a Brzezinski is always worthwhile. But too many of the lecturers we get come in and just talk, ramble on without notes, or tell old "war stories"—not literally, of course, even if it is the National *War* College. The worst are those who come in unprepared and tell us they'd "rather hear from us," that they'd be "happy to answer questions." If I'm serving as the host in such sessions, I always demure politely and say, "Oh no, we'd rather hear from you first and then we'll ask our questions"—if only as a device to reveal how vapid and unprepared the speaker really is. I would say about two-thirds of our speakers come prepared and do a good job, and about one-third do not—too high a proportion.

The interesting thing about these core lectures is that the NWC faculty only rarely delivers them. Most are delivered by outside lecturers (as above) who may or may not speak to the day's subject matter. Now, the *facts* are that quite a number among the civilian faculty at the College are better informed, better read, and better lecturers than the guests who are brought in. They are quite capable of delivering these core lectures themselves. But the military brass has made a decision that we civilian professors on the faculty are too low-ranking for that. In the military's way of thinking based on hierarchy and rank, it is better to have a poor lecturer who holds a high rank than a good lecturer from the faculty who ranks not so high.[8]

Several of us from the civilian faculty—Terry Deibel, Bard O'Neill, Paul Godwin, myself—have on occasion delivered lectures to the entire student body in the core course. In my case, it was the key introductory lecture in the Geostrategic Context part of the core, which provides the concepts and structure for that entire section of the course. Afterwards, several of my students came up to me to tell me how well it had gone—but with an obvious note of surprise in their voices. Even though I think of myself as an experienced lecturer, they were surprised that someone with my low rank—a "mere" professor at NWC—could bring it off.

After the regular morning lecture, which normally takes place four days per week, there is a fifteen-minute break. At 10:00 the students reconvene in smaller groups of twelve to fifteen, presided over by a faculty member, civilian or military, for discussion. The discussion sessions are meant to have the students raise questions about the preceding lecture and discuss the main topic and readings for that day. I have always found it awkward to go into a seminar-size group like this and ask, "Are there any questions?" because that way there never are. Hence, my modus operandi, which I'm sure violated some military regulation somewhere, was to spend the first fifteen to twenty minutes lecturing myself, providing focus, and offering an outline for the discussion—before opening it up for discussion. The students appreciated this approach; I consistently received *the* or among the highest scores in the College on my teaching evaluations.

In these seminars it is often hard—and you have to work at it—to earn the students' trust and confidence. Remember, these are forty-year-old guys;

they've been around; they've seen a lot of the world; they understand bureaucracy. They're often as old and experienced as you are; what do you know that they don't? Actually, most of us on the civilian side do know a lot more than they; we know the literature, the theory, the concepts, and often the situation on the ground in various countries and regions far better than they. But it often taken two or three sessions before this sinks in to the students; when it does, it usually cements our relations with them. The military faculty at the College often have a harder time: frequently, they've not taught in a classroom before; they're at the same rank as the students and, therefore, have no authority over them; and they don't know the literature beyond the day's immediate readings in the way the civilian professors do.

As faculty, we are supposed to evaluate the students on their oral and written work. The oral work consists of their participation in the classroom discussion; the written work consists of a brief, five- to seven-page "thought piece." No original research is required in these papers—we are not training future academics at NWC—and the papers submitted are nowhere near the level of a research paper in a good graduate seminar.

In the evaluations we do for the students, furthermore, no letter grades are used, and we are not allowed to say anything bad about the students that might penalize their careers. Indeed, it is even unlikely that our evaluations become a part of the students' permanent record. So most of us on the faculty develop some meaningless pat phrases—"good work," "nice job," "good participation," "interesting ideas"—that we use to fill the space on the evaluation forms. The result, not unlike the State Department's Foreign Service Institute where I once taught, is that the evaluations we do as faculty are worthless, totally irrelevant to the person's future career. Not only does this eviscerate the performance review process; it also means that some totally unqualified individuals manage to get through the process and go on to do even greater damage at a higher level.

At 11:30 everyone takes a break for lunch. By Washington, to say nothing of European, standards, this is a very early lunch break; but this is the military and that's what they do. Most of the students, since physical exercise is valued in the armed forces more than intellectual exercise, use that opportunity to run, work out, or play baseball or volleyball. Most people do their workouts at noon on the NDU campus; some brave souls go outside the base and do long-distance running in the neighborhood. The faculty are free to join these activities: I often used the lunch break to play tennis, get in a round of golf, or walk around the base (I used to be a runner but, with age, my knee cartilage has worn thin). Most days I would brown-bag, get in a little writing over the midday break, and then go for a stroll around the grounds. Once a week, usually on Fridays, despite the early hour, I would invite my think tank and policy-community friends in for lunch at that comfortable, subsidized, all-the-food-and-wine-you-can-consume officers' club at Fort McNair.

Afternoons at NWC are devoted to what are called "Advanced Seminars." These are more specialized courses, usually for twelve to fifteen students each, on specific countries, regions, or topics. The seminars run for twelve weeks, September to December in the fall and January to March in the spring. While at NWC, I regularly taught a two-course sequence: South America in the fall and Central America and the Caribbean in the spring. Frankly, the sequence—and the syllabus prepared for it—was modeled on the two-semester sequence that I'd long taught in my civilian university background. But it didn't work quite the same way at the War College: whereas at my civilian university I could always count on many students, because of their inherent interests in the subject, taking both halves of the sequence, at the War College that never happened. The military wants to train generalists and that means one segment is enough; it does not want to train specialists in any one area or subject.

Though the course is billed as an "Advanced Seminar," it is actually a long ways from that. It meets once a week, sometimes twice, for an hour-and-a-half—half the length of a serious graduate seminar. *None* of the students has any background in the area. They don't know anything about the issues, the language(s), or the literature. Unlike even an undergraduate course, there are no prerequisites for the course: nothing in Spanish, Portuguese, or comparative politics. Mainly, I found out, they signed up for my seminar because Latin America was the one area of the world they knew nothing about. On the other hand, these are bright guys; they pick things up fast. But the readings, by order of the War College high command, are easy and elementary; there are no tests; and the papers expected are at the same level as those in the core course: short, no research required, pretty elementary. I would say the course, rather than being "advanced," certainly not at a graduate level, is really equivalent to my undergraduate survey course. Better than nothing but not at the level to function adequately in the area at a policy level, which NWC graduates are expected to do.

The official handbook of the War College says the work day runs from 8:30 a.m. to 4:30 p.m. That means there's supposedly time for the students to take two of the one-and-one-half-hour-long "Advanced Seminars" in the afternoon. But the College Executive Officer discovered that few people were signing up for the late-afternoon seminars and, if forced to do so, objected strenuously. That is because of the rush-hour traffic on Route 395, running south out of central Washington into Virginia. Because their past and future assignments are likely to be in Virginia (the Pentagon, Fort Mead, Fort Belvoir, Quantico), most of the officers have homes on this southern side. And after 3:30 the traffic on 395 is horrendous. So the War College relented, did some rescheduling of the afternoon seminars, and accommodated the students' commuting needs, showing again that it is the students who rule at NWC, and not the faculty or central administration. By 3:00, or 3:30 at the latest, the halls of the War College are all but completely deserted as the students flee to beat the rush.

Much the same applies for evenings. Because these are forty-year-olds, almost all our students are married and have children. Therefore, we almost never have evening programs. Once again, the students rule!

NOTES

1. For some reason, perhaps related to the fact it's Bowl season and I'm about to watch one of the games, I'm reminded of the term (and oxymoron) "student athlete." The comparison would be: NWC is to a civilian university graduate program as student athletes are to serious academics. Is that too extreme, unfair? I invite rebuttals.

2. The curriculum described here is the one taught in the early-to-late 1990s when I was at NWC. Since then, it has changed in various particulars.

3. Richard Downie, *Learning from Conflict: The U.S. Military in Vietnam, El Salvador, and the Drug War* (Westport, CT: Greenwood Press, 1995).

4. John Gaddis, *Strategies of Containment* (New York: Oxford University Press, 1982).

5. Since I was at NWC, the military part of the curriculum has been cut back substantially.

6. Howard J. Wiarda, *Foreign Policy without Illusion* (Chicago, IL: Little Brown/Scott Foresman, 1990). See also Wiarda, *American Foreign Policy: Actors and Processes* (New York: Harper Collins, 1996); Wiarda, *The Crisis of American Foreign Policy* (Lanham, MD: Rowman & Littlefield, 2006); and Wiarda, *Divided America on the World Stage* (Washington, DC: Potomac Publishers, 2009).

7. Howard J. Wiarda, *Adventures in Research,* Vol. III (Lincoln, NE: iUniverse, 2006), Chapters 20 and 21.

8. The issue of where the faculty ranks and by what criteria is a controversial one at the College. My view has always been that there is no higher position in the *world* than a full professor with tenure. But the military does not view it this way; more on this issue will follow.

Chapter 4

Students and Faculty

The National War College conducts a senior-level, presumably at the Master's degree level, course of study in national security strategy in order to prepare selected military officers and federal officials for high-level policy, command, and staff responsibilities. The War College especially focuses on national security policy and strategy and emphasizes a joint- and inter-agency perspective.

Reflecting this emphasis, the student body is composed of equal representation, or about 25 percent of each, from the land (Army), sea (including Navy, Marines, and Coast Guard), and air (Air Force) services, with the remaining quarter of the class drawn from the various civilian federal departments and agencies, mainly the State Department and the Central Intelligence Agency. In addition, the student body from the beginning included students from other countries, called international fellows, drawn mostly from NATO ally countries but increasingly from the Third World as well, and now including students from the former Soviet Union and Eastern Europe.[1] The War College began in 1946 with a class of fewer than thirty; during my time there, the class size was set at 180, give or take a couple.

The faculty at the War College during my period there numbered forty-six, divided about equally between the Department of Military Strategy and Operations and the Department of National Security Policy. That number includes the commandant, XO, two deans, and two department heads who also teach classes. The ratio of faculty to students is a low four-to-one, reflecting the fact that this is a high-level graduate program and has no large classes, except the core course lectures, as in an undergraduate college. Of the forty-six, twenty-six during my years at NWC were military and twenty were civilian; fifteen of the forty-six had Ph.D. degrees.

In its early years, with George Kennan and Bernard Brodie there, NWC had a quite distinguished faculty and was a major center for strategic thinking and writing, as well as teaching. Among other things, the containment strategy, NATO, the Marshall Plan, Cold War strategy, the doctrines of massive retaliation and a graduated response all had their origins, in whole or in part, at the College. But then the College went into a twenty- to thirty-year decline during which its faculty was not only not particularly distinguished but very little research and policy writing took place. That began to be reversed in the early 1980s when a series of congressional hearings and investigations revealed the War College's limitations,[2] and then in 1986 when the Goldwater-Nichols Reform Act singled out the College and called for a higher-quality faculty and improved teaching methods.[3] Goldwater-Nichols led to the creation of the Title Ten academic positions at the College (under which I was brought in), which enabled it to pay high salaries in order to attract a higher-quality faculty.

The irony is that, while the Title Ten positions succeeded in attracting better NWC faculty, it also led to rising discontent in faculty ranks. For as the College brought in better faculty, it also brought with them the demands that these faculty would bring with them—for tenure, sabbaticals, research facilities and time, travel money, released time, and greater academic freedom. All of these are, of course, anathema to the military mind and command structure. So, as the College sought to improve its faculty, it also brought with that effort a whole new set of problems that the military institution had not experienced before. That ongoing tension is at the heart of the discussion in this and subsequent chapters.

THE STUDENTS

The students at NWC are all about the same age—early forties. And, therefore, in the military system they are all at the same rank—lieutenant colonel or colonel, or its equivalent. Many of them, about a quarter, are slated for higher ranks—general or admiral, "flag officers"—which is why they were selected to attend the War College by their respective services. Some of them are even awarded their flag officer's star while they are students at the College; when that happens, there is usually a big round of applause, and often the promotion ceremonies are held there as well.[4]

The three-quarters of the class who are military conform to many of the stereotypes we have about the armed forces. They tend to be conservative, religious, and disproportionately from the South and Middle and Small-town America. They are Red Staters for the most part—Republican; they voted overwhelmingly for Ronald Reagan, George H. W. Bush, Robert Dole, George W. Bush, and John McCain—although the percentage voting for "W" dropped off in 2004 and for McCain in 2008. During the 1990s, except for a handful, the students had a positive disdain for Bill Clinton, seeing him as a lying, cheating, draft-dodging, pot-smoking president—all traits that are inimical to the military culture. Most did not favor the further integration of women into the armed forces, and they were positively disdainful of Clinton's efforts for purely domestic political reasons to allow gays into the armed forces.

There are many nice, admirable, even lovable traits in these military officers. They are, for the most part, honest, straight-forward, honorable, patriotic, and above-board. They are energetic, gung ho, and physically fit (part of the military culture). In high school and college, or the military academies, they were bright, hard-working, National Honor Society students but not often valedictorians or salutatorians: B+ rather than A students. They tend not to come from elite private schools but from public schools, colleges and universities where they were often ROTC students, or the academies. They care deeply about their country, are often impatient with politicians whom they see as two-faced, and are not nearly so cynical about the American political system as, for example, I am. They believe in the American Dream because they have lived it, often rising from rural, small-town, and lower-middle-class families, winning scholarships to college or the military academy, and rising in the ranks by dint of energy and hard work to upper-middle-class status. In my years at the War College, a full colonel might earn a salary in the neighborhood of $80,000 a year.

The officers at the War College share the earlier ethos, now sadly in decline in America, of Ben Franklin and Horatio Alger: work hard, save your money, be ambitious, study hard, get ahead. They are energetic, ambitious, and, even though out of school for the preceding seventeen to eighteen years, eager to learn. Because they have been in the military for those seventeen to eighteen years where conversations and written communiqués are often cryptic, or because they have engineering backgrounds, the officers are not good writers, which puts them at a disadvantage compared to the State Department students in the class. But I have found their analytic and critical skills to be as good as the civilian students. Almost all of them are capable of M.A. and maybe Ph.D. work. Moreover, they are in an institution, the armed forces, that is the most admired institution, *by far,* in the country.

The one-quarter of each class who come from civilian agencies do not seem to be governed by the same criteria as the military students. First, the civilians from the State Department and other agencies are often not quite sure why they were selected for NWC. The reasons are not always rational:

maybe their agencies had no jobs for them at the time and so shipped them off to the War College for a year; or maybe their bosses simply wanted to get rid of an unfortunate bureaucrat and thus assigned them to NWC. Second, and related to the first point, after they graduate from NWC, the civilians are not necessarily slated for higher things—an ambassadorship—as are the uniformed students. Many continue to languish in their respective agencies; there is no correlation among the civilian students between attendance at NWC and rapid promotion within their own agency.[5]

Actually, I am not so sure it is all that different for the military students. Although these students are always lauded by the NWC command as "the best and the brightest," I am not convinced that is consistently true. I remember the former dean of faculty who recruited me to NWC telling me how much I would enjoy teaching there because the students are "so bright." They are "wonderful," he said, a "joy to teach." And that is the official view at the College: that the uniformed students are the "cream of the crop," that they were carefully selected by their services, are "comers" and "high flyers," and slated for "higher things"—i.e., a generalship or admiralship. But I did not find them to be such consistently good students, nor am I convinced that the nation and our security are in such good hands with these guys.

First of all, the students themselves, like the civilians, are often selected for the NWC course for reasons other than merit. Often their services have no jobs or billets for them at the time, and so they are shuffled off to the War College just to fill a slot there. Or else, the services may want to get rid of an obstreperous officer, can't "fire" him immediately, and so off to the War College he/she goes.

In my classes and in one-on-one discussions I would often ask my uniformed students why they were there. Many had no idea why their services (remember, it is the services who select the students to attend NWC; the College itself has no "admissions office" and little say in the selection) had selected them; they often said there were other officers more qualified than they. But perhaps their career schedules didn't fit the NWC schedule, or perhaps there were other "billets" that needed to be filled. Many of the students came to NWC as last-minute assignments, in large part because their respective services had no other slot for them at that moment. The point is: there are many reasons students are selected for the NWC course, and not all of them have to do with merit.

Former XO Kirk Lewis had some interesting takes on this. He said that the military students at NWC had forty-year-old bodies but only twenty-year-old minds. By that he meant that these were guys who had been tank drivers, airplane jockeys, or boat pilots in their previous, twenty-year military career; but they had not done anything intellectual in the last twenty years. They often had strong views shaped by their military careers and prejudices, but often little evidence or analysis to back up these opinions. They were not intellectuals, he said; indeed, they were often anti-intellectual. He was less sanguine about the students than had been the dean of faculty.

Students come to the War College already equipped with both general and specialized knowledge. Remember, these are forty-year-old guys with nearly twenty years of military experience. As lieutenant colonels or colonels who have navigated the often mysterious and labyrinthian corridors of the Pentagon, DOD, and their own service's promotion process, there's not much we can teach them about politics or bureaucracy. In addition to this general knowledge, they often have specialized knowledge concerning tank maneuvers, aerial dogfights, the occupation of Panama, or the Iraq desert war.

What they need at NWC, therefore, is an integrating curriculum that they can all use and be helped by, on international relations, foreign policy, the diverse countries and regions of the world, and the relationships of policy, diplomacy, and armed might. They need a common set of understandings heretofore lacking because most of them do not have prior training in IR and foreign policy. One of NWC's "star" graduates, Colin Powell, put it well. He said he'd lived his entire life within the narrow cocoon of the Army and, even more narrowly, in his particular specialization. Then he came to NWC, met people from other services and other agencies, heard a whole new set of IR and foreign policy ideas, and enormously broadened his perspective. That, he says in his memoir,[6] is what prepared him for the White House, the National Security Council, chairmanship of the Joint Chiefs, and the position of Secretary of State.[7]

Surprisingly, I found the students to be far less ambitious than I thought they would be. My assumption going in was that they were all high flyers—rather like Tom Cruise in *Top Gun*. And we did have our share—more than our share—of guys who fit that stereotype: cocky, hot-doggers, supremely self-confident to the point of arrogance. I had also assumed that their ambition was to be like Brent Scowcroft or Colin Powell: military officers who had used their international relations training at NWC to go on to bigger things, such as a Joint Chiefs chair, national security adviser, or Secretary of State.

But no, that was not the case. Most of my students were not that ambitious. Nor were they very intellectual. They did not want necessarily to be among the Scowcrofts and Powells of this world. They did not want to get Ph.D.s in international relations, as Scowcroft had, at Columbia, no less. Most of my students saw the year at NWC as a nice interlude, but they were content to go back to being boat captains, tank drivers, and air jockeys. They did not want necessarily to wed their military training with academic prowess, nor to play the political game (Powell, Wes Clark, Barry McCaffree, Jim Jones, et al.) and go into a high civilian slot. Most wanted less ambitiously to get on with their military careers and retire comfortably in the not-too-distant future.

Nevertheless, we as faculty were constantly assaulted with the insistence of how good the students were. And indeed the College was certainly run for the students and not the faculty. The students were lauded, while the faculty was only there to serve them. It was the most curious "college" I've ever been associated with. The students were praised *above the faculty.* Students were showered with plaudits, the faculty was demeaned. At every turn, it was emphasized that the College was there to help and serve the students, not the other way around. In this context, it would be easy to see why the College would not value the research or writing of its faculty, would not favor or grant tenure to its faculty, would see the faculty, in the words of one commandant, as a "pass-through" faculty: here today, gone tomorrow, without permanence, meant only to enhance the careers of the students.

Kennan's and Brodie's original conception of a prestigious faculty producing cutting-edge research on national security issues had thus been completely abandoned. One of the deans told me he thought that the research function ought to be at the nation's prestigious universities, not at the War College. But by that time (the '90s) foreign policy studies at such prestigious universities as Harvard and its Center for International Affairs (CFIA) were being downgraded or abandoned in favor of narrow, arcane, political science research based on "data sets." As the field of international relations had presumed to take on "scientific" qualities, it had largely abandoned the search for "good policy." In fact, even as the dean cited above spoke, research and writing on serious foreign policy issues had shifted from the nation's universities to the Washington-based think tanks. Though that is getting ahead of our study a little, it is in keeping with the overall thesis of this series: the shift in power and foreign policy influence since the 1960s away from our universities and toward the Washington think tanks, policy "shops," and lobbying organizations.[8]

Students at the War College were often pampered, told how good they were, and treated deferentially by the commandant, XO, and deans. We, as faculty, were frequently told to treat the students with respect (even if they failed in their assignments), not to criticize them (even if they said stupid things in class), and to consider them as our "equals" (hard to do in a classroom when you're the teacher and they're the students). I attributed this deference and what my father would call cow-towing to the students to very pragmatic considerations: the College wanted the students to have a good impression of it; the College wanted the students to be supportive of the program so as to make it more attractive to the next year's classes; and, since many of NWC's graduates would soon be flag officers, they would soon be in command positions to affect NWC's budget, reputation, and standing in DOD and with the Joint Chiefs.

There are numerous instances I could cite, large and small, of special favoritism to the NWC students, often at the expense of the faculty. When I arrived at the War College, such favoritism and cow-towing seemed (and is!) very strange, and it took me a few weeks to figure it out. Because in my civilian university I would *never* cow-tow to students, and the idea that they were equal (sorry, former students) was laughable. But not at the War College where the students were not only our equals but treated as though they were superior. NWC is *overly* student-oriented. As my former faculty colleague Ron Tammen once commented, "When the students complain to the administration, things happen right away; when the faculty needs something, a committee is formed."

Here are some illustrations of this favoritism. Once when I was talking to the commandant about the possibility of getting my CSIS friend and colleague (and former Jimmy Carter National Security Advisor) Zbigniew Brzezinski to come to the College as a one-year visiting professor (The Kennan Chair), and the commandant wanted me to sound out Zbig about the prospect, he remarked, "Well, he would come here because of the reputation of the student body; the faculty would not attract him here." Well, gee, thanks, commandant, for your expression of confidence in your faculty! And, by the way, Zbig *was* interested in the position but could not do it that year; instead, we got future Deputy Defense Secretary Paul Wolfowitz who proved to be a disaster (arrogant, unfriendly, seldom present) as the Kennan Chair.

On another occasion, the faculty was told that, henceforth, the student discussion sections would be reconstituted every four weeks instead of the previous eight. Thinking like an academic and in terms of pedagogy, it's hard to get to know students and evaluate their progress in eight weeks, let alone four. Now we would have a whole new slate of students to evaluate every four weeks. This was bad pedagogically, but that had nothing to do with the change. The reason the College implemented the change, over the faculty's objections, was that this way the students would bond as a class and get to know each other better. Bonding had clearly trumped teaching and pedagogy.

Another time, the commandant explained to us his ideas of pedagogy—even though he had no experience as a teacher or educator. He viewed lectures as "passive time" and all but worthless: you sit back and you listen. But even though you're presumably gaining new information and insight from the lectures, he thought this was a waste of time; if he had his way, lectures (with real expertise) would be abolished. Instead, there would be only "active learning"—discussion sections. These afforded the students a chance to speak up and air their views—even if they had little worthwhile to say. Faculty member Paul Godwin told me once that he had difficulty staying alert during these student discussions because he'd heard the same simplistic arguments so many times before. But not so for the commandant; he did not value real (faculty) knowledge and expertise but only the *exercise* of having the students speak up. Once again, the College exhibited misplaced priorities.

And then came the great book heist caper. One day when I was out of my office for a brief time, a student came by, saw the office was empty, came in, and "borrowed" one of my books off the shelf. At least he had the courtesy to leave a note. But in my academic (or any other!) world, you *never* enter another person's office uninvited, doubly-never when the occupant is not there, and you certainly don't "borrow" someone else's property without permission. In addition, this was one of my own privately owned books, and I was using that book at the time for some of my own writing.

I was angry, went storming off to see the dean, and insisted that the student not only bring the book back but also be punished. The issue became a big deal at the College and went all the way up to the XO and then the commandant. It's hard to believe, but they actually sided with the student. From their point of view: (1) my office is not sacrosanct, but is only a convenience to serve the mission of the College; (2) the student, therefore, had a right to enter, even without me there; and (3) any books that were there ought to be for student use and not my own. Wow! To me the issue was clear-cut: the student had entered my office without permission and had taken my property; but to NWC my interest as a faculty member was subordinated to a higher College and student interest. Talk about misplaced priorities.

Shortly thereafter, a not dissimilar case came up. This time a popular student, president of the Tailhook Association (which arranges social events for Naval officers and their friends—often in Las Vegas), was caught cheating. He turned in the same paper to two classes, a no-no and a clear violation of the NWC honor code. Since the paper was turned in for my class initially, I was involved. Since any little negative mark in your file is enough to ruin an officer's career, this became a big deal at the War College.

At a civilian university, the response to this kind of violation would likely be low key: talk to the student, force him to write a new paper, tell him to drop the class, or maybe fail him. But in the military, in this litigious age and when professional careers are on the line, things can quickly escalate. Two lawyers from the judge advocate's office came to interview me; they wanted not just the facts but were also soliciting my opinion on the context and how this could be resolved. The accused student had done well in my class; he was remorseful; and the JAG lawyers plainly did not want to ruin his career.

The options seemed to be: (1) have him write the paper over; (2) refuse to allow him to graduate; or (3) drum him out of the Navy. The commandant would decide.

Here's another, this time minor issue. When we had guest speakers whom I knew personally and the schedule was propitious, I'd often invite them to stay for lunch at the Officers' Club. And then invite several of my faculty colleagues to join us—Dutch treat. That's how you'd do it in academia; right? But not at the War College—and for the same reasons suggested above. When the College brass heard of these lunches, they objected. They wanted the students invited. Well, OK, one or two; but soon the students, at NWC's insistence, were in the majority. No thanks; I see enough of my students in class and don't need to see them over lunch, too. Pretty soon I and other faculty stopped going; shortly after that, the program died.

I would say that NWC was skirting right on the edge of pandering to its students. Everything seemed to depend on whether the *students* were happy and satisfied. The reason for this, I was told by one of the deans, is that it's the students who, if unhappy for any reason, complain to the Joint Chiefs. The faculty cannot do that. And since NWC is a creature of the Chiefs, it has to be careful of its political relations in this regard. And we faculty and administrators have to make this a good experience for them. Otherwise, the budget will be reduced and the College closed down. I understand that logic, but it's still no excuse for pandering to the students.

The longer I stayed at NWC, the worse (from my point of view; "better" from the students' viewpoint) the situation seemed to be becoming. *Everything* seemed to be done from the point of view of the students' best interest, not that of the College and certainly not that of the faculty. The faculty was now actively discouraged from engaging in research (time away from the students) and publication. The president of the student body was given his own personal parking space, closer to the building than any of the faculty. He also had his own privileged reserved chair in the auditorium. Meanwhile, I was told by my department head and dean to allow my students to skip class to join the bowling team. Bowling? The logic was that bowling builds camaraderie and class unity—more "bonding." But in precisely that same week a faculty colleague with real expertise on the issue was denied the opportunity to be away for half a day to brief the military command on EU and NATO enlargement into Eastern Europe. More misplaced priorities.

As I was preparing to leave NWC in 1995–96, plans were underway at the College to make the program even more student-centered than it already was. All kinds of plans were being discussed at the central-administration level to exercise greater central control over the curriculum—and to exclude faculty from participating in this process. The plans included having the faculty teach more, offer more sections, teach more students, and do more administrative (paperwork!) things. The core curriculum would be reduced and the students would design their own courses, according to *their* needs, and at the times *they* desired. As one uniformed NWC administrator put it at the time, "There is only one reason for our existence, and that is to serve the students."

These proposals fit in with other plans then coursing through the College. They came not just from the NWC commandant but also from the NDU president, and even the Joint Chiefs. The idea was to have the faculty work harder and teach more; but the greater effort would all be dedicated to teaching and none to research. The Chiefs reasoned that they had their own in-house, DOD policy planning staff and think tank, to say nothing of the private RAND Corporation, the Center for Naval Analysis (CAN), and NDU's own INSS, so why should the NWC faculty also do research? These proposals were never fully implemented, but the fact that they were advanced and seriously discussed shows an appalling lack of understanding by the military brass about teaching, especially at this high level, and the intimate relations between research and good teaching. Thank goodness I was able to leave the College within a few months.

THE FACULTY

Remember those figures offered at the outset: NWC had a faculty of forty-six persons, twenty-six of whom were military and twenty civilian, fifteen with Ph.D.s. The faculty was divided into two departments: a Department of Military Affairs and Operations (DMAO) and a Department of National Security Policy (DNSP). I was in the Department of National Security Policy; my official title was "Professor of National Security Policy."

Almost all the military officers on the faculty were located in the Department of Military Affairs and Operations, although DMAO did have a handful of civilian faculty from CIA, the Defense Intelligence Agency (DIA), the Office of the Secretary of Defense, and other agencies. I had relatively little contact with DMAO: it's not that we were hostile, it's just that I was in a different department and did not teach the Foundations of Military Thought and Strategy or Military Strategy and Operations parts of the core curriculum for which DMAO was responsible. In contrast, almost all the faculty in the Department of National Security Policy were civilian, although we did have a handful of military officers in our department; and almost every civilian faculty member within DNSP had a Ph.D. You can probably guess that this division would provide the basis for almost constant tension and problems between the two departments.

Since I was not a member of DMAO, I won't say much about that department's faculty. For the most part, I found them to be energetic, enthused, and good guys—not unlike the general student body. How they were chosen to teach at NWC was something of a mystery: generally, they were selected by their own services to fill this "billet," though the War College often tried to have a say in their selection as well. Most of them were inexperienced and sometimes neophytes in teaching—in keeping with the commandant's oft-expressed sentiments that teaching experience was unnecessary because "every officer is a teacher." And because these were, for the most part, serving military officers who would only be at the College for two or three years of their lives before going on to other assignments, they had no interest in, or incentive for, engaging in research or publication. No, DMAO was purely a teaching department consisting almost exclusively of military officers. Many of us in the other department suspected that was the model the commandant and the NDU administration wished to impose on the entire college.

The other department, National Security Policy, was more diverse—and also more like an academic department in a civilian university. It consisted of about two-thirds civilian academics with Ph.D.s, with a scattering of persons who may or may not have had doctorates or law degrees, from the State Department, CIA, or other agencies; and the rest military, at the colonel level. There was frequently a problem with the military officers assigned to the Department: they had little or no teaching experience and no specialized knowledge base. In that case, those of us on the civilian side with long teaching experience were supposed to take them under our wing, let them sit in on our own classrooms, and teach them how to be good teachers. But how can you be a good teacher if you lack an independent base of knowledge? In addition, and still thinking like a civilian academic, I saw this mentoring of new and inexperienced teachers as adding to my teaching load and not part of my contract. Raising these objections publicly, I was accused of losing sight of the College's mission and not being a team player.

The Goldwater-Nichols military reform act of 1986 had recognized that the civilian teaching faculty at NWC had been allowed to deteriorate, was not very good, and badly needed to be improved. To that end, Goldwater-Nichols had called for higher pay and better working conditions for the civilian faculty and the recruitment of a top-flight faculty. It was under those conditions that I'd been recruited in 1991 and others then and in succeeding years. We'd been brought in as what were called Title Tens, a special category outside the regular General Services (GS) categories that enabled DOD to pay us more and provide better working conditions. My pay was certainly quite adequate, about 40 percent more than I was then making in my civilian university. For the record, I was told by the dean that I was the highest paid faculty member at the University. That was frequently thrown in my face whenever I complained about anything.

Over the course of the decade after Goldwater-Nichols, from the mid-1980s to the mid-1990s, NWC had, in fact, recruited a really good faculty. I would say it was as good as all but one or two (Harvard's CFIA in the 1980s[9]) civilian university international relations contingents in the country. The faculty included:

Janet Breslin, Ph.D., specialist in Congress
Hugh De Santis, Ph.D., Chicago, specialist in Europe
Terry Deibel, Ph.D., Fletcher School, foreign policy
Paul Godwin, Ph.D., Minnesota, East Asia
Mel Goodman, Ph.D., Indiana, Soviet Union
Dale Herspring, Ph.D., Southern California, Eastern Europe
Richard Melanson, Ph.D., Johns Hopkins, foreign policy
Bard O'Neill, Ph.D., Denver, Middle East
Marvin Ott, Ph.D., Johns Hopkins, Southeast Asia
Al Pierce, Ph.D., Tufts, military affairs
Thomas Schelling (visiting), Ph.D., Harvard, strategy
Charles Stevenson, Ph.D., Harvard, Congress
Ronald Tammen, Ph.D., Michigan, Congress
Cynthia Watson, Ph.D., Notre Dame, Latin America, Central Asia
Howard Wiarda, Ph.D., Florida, Latin America
Paul Wolfowitz, Ph.D., Cornell, security affairs

Well, there you have it. Quite a stellar faculty. All Ph.D.s; all with degrees from quite good places. If one considers only the comparative politics, IR, foreign policy faculties, this was as good a group as almost anywhere in the country.

A number of features of this faculty stand out:

- About an equal number came from public universities as from elite private or Ivy League universities.
- Almost all the faculty came out of—and this was done purposely by NWC—prior government service, working previously either with the State Department, Congress, or other executive agencies. I was the first faculty member appointed directly out of an academic career background, and even then I was told that it was my prior State Department, think tank, and Washington experience that got me the job.
- The faculty was overwhelmingly male; there was only one woman, eventually two.
- The faculty was almost all the same age: forties, comparable to the students. That was also done purposely.

A unique and curious feature of the DNSP faculty is that we were all of one rank. We were all "Professors of National Security Policy"; there were no Assistant, Associate, and Full Professor ranks, although there were distinctions in terms of pay. The absence of academic rank not only made us all "equal" to each other, it also made us of equal rank, supposedly, to the colonels in the other department and to the colonels and other non-academics in our own department. And, frankly, from the Army's point of view, a system of academic ranks that overlapped with and sometimes contradicted the system of military ranks as well as DOD's GS categories for civilians was just too messy and complicated to deal with. However, even though the College had no formal academic ranks, informally, in terms of prior experience, publications, and academic record, we all recognized that there are differences and different standings in the scholarly or academic pecking order.[10]

The issue of ranks, in academia and the military, is extremely complicated; I cannot resist telling one story about that here. Quite frankly, I think a tenured full-professorship in a good university is about the highest rank in the world, but for protocol, administrative and even disciplinary reasons, the Defense Department classifies professors at about a GS-15 or colonel level. However, because of my previous think tank experience, I am friends and sometimes socialize with the likes of Jeanne Kirkpatrick, Henry Kissinger, Brent Scowcroft, Richard Cheney, Zbigniew Brzezinski, Madeleine Albright, and others of similar high social or political standing.

One time I was at a cocktail/buffet reception at that level and ran into the War College's commandant (a two-star general) there. He was obviously stunned to see me. He didn't say anything then, but the next morning he called me into his office and accused me of "socializing above my rank." I almost fell over; I'd never heard of such a thing before. Not only do I have this elevated view of professors, but in my line of work I'm used to interviewing presidents, prime ministers, and foreign ministers, as well as having those A-list friends listed above. And who is this commandant, anyway, to accuse me of socializing above my rank? I just about told the commandant to stuff it; I did tell him I would continue to socialize with whomever I wished and that it was all none of his business. But from a military point of view, one can understand what he was thinking and that, for the good of order and discipline, you should socialize only with those of your own rank: enlistees with other enlisted men, majors with majors, colonels with colonels, generals with generals, etc. But where do academics fit into this ranking system? In the following chapters we will have much more to say on this issue of the clashing cultures, military and civilian-academic, at the War College.

I would not say there was bad blood, but there was certainly a lot of tension between the military and civilian faculty and their respective departments at NWC. The military guys tended to resent the civilians because we were paid more than they, could travel more, had more freedom and more free time, could read books and engage in research, and so on. At the same time, the civilian faculty tended to look down on the military faculty because they lacked Ph.D.s, had slim teaching credentials, had not gone through the rigors of passing comprehensive exams ("comps"), writing a dissertation, publishing, or earning tenure, as we had. They thought we were overpaid; we thought they were not only overpaid but unqualified as well, at least in terms of having the credentials to teach at the National War College. Most of these tensions were kept under wraps, and we were cordial to each other, but below the surface a lot of resentments were brewing.

There were also resentments and bad blood between and among the civilian faculty at NWC. Most of the civilian faculty had come out of prior government service—State Department, Congress, CIA—which meant that, as full-time civil servants, they generally had weak or nonexistent research and publication records. In quite a few cases, they had gone into government service precisely because, even though they had the requisite Ph.D. degree, they were poor writers and were unable to get their research published. They tended, however, to be dismissive of those of us who had come out of academia but had more limited government experience. On the other hand, the "academics" had longer and stronger research and publication records, which often rankled with those who did not.

I was one of those who came to NWC with a strong research record. And I am very much research and publication-oriented; once at NWC, I continued to push for more time and resources to be devoted to policy-oriented research and writing—the Kennan model. But then the "old-timers," those with long government service before coming to NWC, got hold of me and told me to tone it down, that the teaching load at NWC was so heavy that I wouldn't have time to do research and write. But I never found that to be the case. True, during the segments of the Core Course that we were responsible for, most mornings, Monday through Thursday, were pretty well taken up; and for two hours on two days a week, our afternoons were likewise devoted to advanced courses or seminar teaching. But that still left Friday morning, three afternoons per week, summers, and large blocs of several weeks at a time when the Military Affairs Department handled the Core. That provided plenty of time for research and writing—by my estimate, about as much as I'd had in my civilian university.

So I continued to do my research and writing, grinding out a succession of articles, papers, op-eds, and at least one book per year. This caused bad blood among quite a number of my non-producing departmental colleagues. First, it made them look bad by comparison. Second, it put the lie to the notion they'd long advanced to the deans and others that they were so busy teaching and with administrative work (grading, evaluations, curriculum planning, etc.) that they had no time left for research and publication. Hence, the more I published, the more it was resented by other faculty. My extensive publication record did not endear me to my colleagues, both civilian and military. Eventually, some among my civilian colleagues began to complain to the dean that, because I was publishing so much, I must be shortchanging my teaching—even though my teaching evaluations were also better than theirs. Some of my (jealous?) colleagues even went so far as to pressure the department head and dean not to allow the secretaries to type my materials, or to have someone monitor my time and computer to ensure I was devoting sufficient energy to my teaching. In these respects, NWC was a little hot bed of jealousies, envy, and petty back-stabbing, to say nothing of, at one stretch, having the department head "spying" on his faculty.

Nevertheless, the College did not explode and, in general and on most occasions, we got along pretty well. The College continued to function and to function more or less effectively; the post-course surveys we did indicated the vast majority of the students, upwards of 90 percent, were very happy with the program and, especially, the teaching at NWC. I would say that the bickering and petty infighting described here was no worse, but no better, than had been my experience at other civilian universities, think tanks, and research centers. The difference was that at NWC we were training the future military and civilian leaders of the country and, therefore, the stakes were much higher. Somehow we managed to hold all these subsurface tensions sufficiently in check to fulfill our teaching and training core mission; and judging from the student responses we did a pretty good job at it.

NOTES

1. Some NWC briefings and discussions are classified, requiring a certain level of security clearance on the part of students and faculty who attend. On occasion, this has been a problem with the Third World and former Soviet bloc students.

2. For the recommended changes, see Committee on Government Affairs, U.S. Congress, "Management of the Department of Defense" (March 23, 1983); Committee on Armed Services, U.S. Senate, Staff Report, "Defense Organization, The Need for Change" (Washington: Government Printing Office, 1985).

3. Committee on Armed Services, U.S. House of Representatives Bill Nichols "Department of Defense Reorganization Act of 1986" (Washington: GPO, 1986).

4. In recent years, only about one-fourth of the graduates made it to flag-officer rank, a considerable decline since earlier decades.

5. With the termination of the State Department's Senior Course, NWC became more of a plumb "Senior Training" assignment.

6. Colin Powell, *My American Journey* (New York: Random, 1995).

7. Actually, I was told by other faculty members, since Powell was at the War College before my time there, that he slept through most of his classes. So much for the idea of NWC preparing future high-level security foreign policy officials.

8. I have prepared a series of books on these themes; see Howard J. Wiarda, *Conservative Brain Trust: The Rise, Fall, and Rise Again of the American Enterprise Institute (AEI)* (Lanham, MD: Rowman & Littlefield, 2009); *Think Tanks and Foreign Policy* (Lanham, MD: Rowman & Littlefield, 2009); *Harvard and the Weatherhead Center for International Affairs* (Lanham, MD: Rowman & Littlefield, 2009); and *Universities in Decline* (Lanham, MD: Rowman & Littlefield, 2010).

9. Howard J. Wiarda, *Harvard and the Weatherhead Center for International Affairs.*

10. About five years ago, NWC created a formal academic title process that created ranks among faculty members.

Chapter 5

Traveling for NWC—and for Myself

One of the perks of being at the National War College was that I got to travel a lot. Some of this was domestic travel; some of it, international

NWC had travel funds, mainly from the NWC Foundation, which I periodically tapped into. They were happy to provide the money so long as the request was placed in national security terms. And that, of course, is what I do as a teacher and researcher: I work on foreign policy and national security issues. As long as the War College was willing to supply the funds, I was willing to travel. On that basis, during my five-year stint at NWC, I was able to travel to Russia twice, Europe four times, Asia three times, and Latin America on numerous occasions. It was NWC that, in considerable part, helped my transition from a scholar with regional expertise to one with genuinely global experience. And that was reflected in my books: away from area studies and to a genuinely cross-cultural, comparative, and global perspective.[1]

If you're a scholar of international affairs, having travel funds available from the Department of Defense's considerable budget and the NWC Foundation is a pretty nice deal. I was happy to take their money! In addition, when you travel under DOD auspices and have a top-secret security clearance as I had, other perks and facilities open up to you: a fleet of military aircraft, helicopters that take you to the remote corners of strife-torn countries, a string of bases that circle the globe, access to NATO facilities and U.S. military attachés abroad, access to foreign militaries because you're with NWC and DOD, and a general sense that you're an important person and in a prestigious position because you're at NWC. Naturally, I took maximum advantage of all these opportunities. And in that capacity, from my position at the U.S. National War College, I saw things and did things that I would never had had access to as a "mere" civilian professor.

Being at NWC and in DOD is not all sweetness and opportunities, however. In the military, you are also frequently "tasked" (a polite word for the old-fashioned "ordered") to carry out certain responsibilities. Often these go above and beyond your contractual obligations. In my case, the extracurricular "tasking" that I received was to lecture periodically on my areas of specialization to Reserve Officer and other training programs held at U.S. bases around the country. Initially, I was resentful of such "taskings" because I thought they took time away from my research and writing and seemed to me to involve extra work for no extra compensation. But over time I learned that these reserve officers and other groups to whom I lectured are wonderful guys to work with and constitute enormously appreciative audiences. After all, for them, a posting abroad is often a life-and-death situation—witness all those reserve units in dangerous Iraq and Afghanistan—and they are, therefore, *eager* to learn about the foreign cultures and countries to which they may be sent. In addition, these programs are held on military bases around the country, enabling me to see, for the first time since World War II, the inside of these bases, how plush they have become, and the country-club-like facilities and atmosphere that exist there. Pretty quickly I stopped complaining about the extra assignments and determined that lecturing at these bases, for the reasons explained above, was an experience well worth having.

We begin with these domestic excursions to a variety of U.S. bases; in the latter part of the chapter, we explore the foreign travel.

DOMESTIC TRAVEL

Van Den Berg Air Force Base, California

My first trip for the National War College was undertaken in October 1991, barely two months after I'd joined the NWC faculty. Our destination was Van Den Berg Air Force Base (named not after the senator, Arthur, but after Air Force General Hoyt), California; the purpose of our trip was to lecture in the RCNSC (Reserve Component National Security Course) on different areas of the world, as part of the effort to update reserve officers' units on current security and foreign policy issues. The RCNSC at Van Den Berg brought in officers from all over the West and West Coast; they would take the briefings we provided and go back and brief their own, home-town reserve units. Also with me on that plane ride were Bard O'Neill and Roy Stafford, experts, respectively, in the Middle East and Europe/NATO. We flew out of Dulles and into LAX; from there we took a prop up to Santa Maria, the nearest civilian airport to Van Den Berg.

Also on the plane out was Madeleine Albright, the future Secretary of State and a friend (not close) from the Washington policy/think tank world. She was on her way to Los Angeles to lecture to the LA World Affairs Council. We chatted for a time during the long flight; she explaining that she was working for a Democrat presidential victory in 1992 and me saying I was part of the George H. W. Bush presidential foreign policy advisory team. That settled, she told me that she, a Czech by birth, had just escorted Czech President Vaclav Havel the evening before to a Georgetown nightclub. Havel likes that kind of action; Madeleine told me that was the first time she'd been in a nightclub in twenty years.

Madeleine told me that her World Affairs Council speech was based on a recent, fifteen-nation survey of European attitudes toward the future: politics, culture, economics, international affairs. As an aside, she told me this was the first serious research (Madeleine had been a Zbigniew Brzezinski Ph.D. student at Columbia) she'd done in years. She also told me she *loved* the political limelight. She said she *loved* being on TV and she *loved* being a presidential adviser (Dukakis in 1988, Clinton in 1992). She said also that in these forums it helped to be a woman—because there were few of them specializing in foreign policy and national security affairs—and that on these occasions she always wore a bright red dress to stand out from all the males in their dark suits on these panels. So that's what big presidential foreign policy advisers talk and gossip about in their off moments!

Flying up to Santa Maria from LAX, we pass over Santa Monica and then Santa Barbara. Both Michael Jackson and Ronald Reagan had their ranches in the mountains up here. We tuck into the valley where Santa Maria is located; Van Den Berg is on the left, occupying a beautiful promontory stretching out into the Pacific. This is some of the most beautiful and least-developed land on the entire California coast. Like the Marine base at waterfront Camp Pendleton farther south, the real estate vultures can't wait, post-Cold War, for these bases to be shuttered as part of the expected "peace dividend" so they can be sold as "beach-front" properties.

I rent a car at the airport and drive leisurely to the base fifteen miles away. With my NWC DOD ID, I go speedily through the gate. This is the first time I've been on a military base in over four decades. I'm in for a shock! My recollection of military bases is of Quonset huts, run-down facilities, and Spartan quarters. But Van Den Berg is like an idyllic, self-contained city. It has everything: its own subsidized shopping center, subsidized supermarket, movie theater, bowling alley, youth center, day-care center, fire department, police, swimming pools, tennis courts, golf courses (plural!), several categories of housing, oceanfront living for the officer corps, etc. Deer and other animals roam the grounds; the flowers are Southern-California lovely; lawns are professionally manicured. It appears I've missed something in military base development and modernization in the last forty years. It's like a plush suburb; in fact, more like a gated (which it is) community.

This is an eye-opener for me. I drive around for a while inside the base before checking in. The bases are not only self-contained; they are immunized against the outside world. Security fences (even tighter now, post-9/11) surround the base; it's almost impossible to get in; and base families rarely leave the compound. On base, there is no crime, vandalism, dirt, pollution, or drugs. For career military officers and their families it's like a cocoon: you're isolated from the outside world; the outside world is isolated from you; all your social life is on the base. Solidly Republican, conservative, and church-going, these bases and their personnel are aghast at the crime, drugs, and general slovenliness of the outside world (including its politicians) and form a culture and society all their own. It's a sheltered and an artificial life, isolated from the outside world, a society within a society, almost a country-club atmosphere. And all this on some of the most beautiful, ocean-front real estate in the world.

My next shock is when I check into the so-called Bachelor Officers Quarters, or BOQ. Again, I expected Spartan conditions: a cot, a footlocker, and a bathroom "down the hall." But this BOQ is plush, more like a modern hotel than a dormitory. The room is complete with kitchen, TV, microwave, refrigerator, and a sitting room—a suite rather than a room. The service is invariably polite, gracious. Across the street is the Officers' Club—bars, entertainment centers, banquet facilities, several dining rooms, wine-tasting, an all-you-can-eat buffet lunch for $4.95. Definitely not a fast-food snack bar! At night, however, at the taxpayers' expense, we go to the Hitching Post Restaurant in the nearby town of Casnalia: maybe the best steak and good Santa Barbara wine I've ever had.

After all this, my lecture and seminar the next day on Latin America are almost anti-climactic. The RCNSC is very well organized, with military precision and discipline; a full colonel is in charge. The audience consists of 180 reserve officers; they're mature guys, with families and already-established careers, not much different from my NWC students. They're very attentive and the questions are probing and well-informed, on Cuba, Venezuela, Mexico, Brazil, Central America, and drugs. Unlike teaching undergraduates, these guys really *want* to know. After all, as in Iraq and Afghanistan where we've heavily used reserve units, these issues may be life-and-death for these guys in coming years. After the formal presentation to the big group, I met in smaller seminar or discussion sessions with groups of twenty or so. They're eager to learn and, to my surprise because I don't expect that from most audiences, enormously grateful that I would come all the way from *Washington, D.C.*, and the *National War College* to lecture to their group.

Wow! A rewarding experience. And the base, a real eye-opener. Are all military bases as plush as this? I was about to find out.

Pensacola Naval Air Station, Florida

Two-and-a-half months after the California trip, NWC "tasked" me to do a similar lecture, what we came to call a "dog-and-pony show," at the Pensacola Naval Air Station in Florida. This was again a program for reserve officers brought together from all over the east and south, just as the earlier gathering at Van Den Berg had brought together the western reserve officers.

On the morning I left Washington for Pensacola, we had just finished the Geostrategic Context bloc in the War College's year-long academic program. My students had given me a rousing cheer as the course ended; four of them came up to me afterwards to tell me how much they'd enjoyed the class. As a teacher, it's nice to have that kind of send-off as you end a course. And even though I'm very critical of NWC at many levels, I always try to do a good job with all my courses.

I flew commercial to Charlotte, North Carolina, then another hop by commuter plane to Pensacola. By the time we drop out of the clouds, we're already over the Gulf of Mexico. We turn, fly over Fort Walton Beach and the plush houses and golf courses below, and bank into the Pensacola Airport. I'm thinking, from the air, that this would not be a bad place to retire: sun, beaches, golf, nice houses, the base PX, and the University of West Florida nearby.

Out of the airport in my rental car, past Pensacola Junior College, down 9th Avenue to Fairfield, and Fairfield to the base. Like Van Den Berg, it's on a gorgeous peninsula overlooking the water, an historic and strategically important base: the Spanish, French, British, and Andrew Jackson in the War of 1812 were all here. There are small islands and sandbars offshore. A lovely setting, the first thing you see on the base are not one but *two* golf courses. The center of the base is all red-brick, colonial architecture; it looks like a college campus. As at Van Den Berg, this is tremendously valuable ocean-front land; the base has its own private beach with pure white sand. The housing is beautiful with green lawns, all manicured, and bougainvillea flowering in front.

The BOQ here may be even more elegant than in California. It's an all-brick, plush hotel. The guest rooms have extra sitting rooms, cable TV, HBO, microwaves, all-carpeted, refrigerators, and a gift package awaiting each guest. Nearby is medical care, security, bowling, and the base tennis courts.

It's like a self-contained city. The nearby Mustin Beach Officers' Club is similarly swank, red brick, looking onto and with spectacular views of the Gulf. Tonight is "Beef and Burgundy" night at the OC: all the beef you can consume and a full carafe of burgundy to go with it. The room is *packed:* lots of base families as well as guests and retirees. Everyone is gorging on the superb beef; it's all subsidized by the taxpayers.

All this is part of a world that neither I nor most other Americans have ever seen. It's closed off to the public—and from public oversight. The logic is, especially in the era of the all-volunteer military, that you need to make the military career and life attractive in order to retain officers and enlisted men. But do we need to go so far as to turn all these bases into plush country clubs? I had always thought of military service as involving just that: service and some degree of sacrifice. Apparently not.

This string of military bases, hundreds of them, at home and abroad, is a closed, "gated" community almost entirely cut off from the rest of the country. There is (almost) no crime, no unemployment, no homelessness, no drugs. It's insular, private, well-guarded—a separate culture within the larger culture. One could go an entire lifetime if you're an American civilian and never know about this military culture; similarly, if you're in the armed forces, your life is insulated from the outside world. It's a complete subsystem unto itself, involving not just multiple bases in each state but abroad as well. As I also traveled outside the country for the U.S. military, I learned that one could spend an entire professional career, and retirement as well, inside the cocoon of these military bases, never learning the language of the countries where one is stationed and never assimilating into the local culture and society. It's a completely separate existence.

My lecture to the reserve officers is scheduled for 8:00 a.m. After the shock of exploring this expansive base and its more-than-comfortable living conditions, the lecture is almost an afterthought. It goes very well and is well-received. The early morning talk means I have the rest of the day, until the 4:00 small seminars, free. In the next couple of hours, I wrote (1) a prospectus for a new book I'm planning on Latin America and (2) a seven-page teaching aid for all NWC faculty for the next part of the Core Course. Tiring of working so hard when I'm in this lush, tropical resort, at 11:30 I head out to explore Pensacola.

There's lots of history here. Pensacola Bay was explored as early as 1519. There are French, Spanish, British, native Indian, and, of course, American influences. The town is quite shabby in its downtown areas as compared to the base. I go out Barrancas Boulevard, which is also the name of the ancient Spanish port. The street names are, alternately, Spanish and English. I proceed along the Bayfront Parkway and the Gulf waterfront. Over the Pensacola

Bay Bridge and into Route 98 to Gulf Breeze. Then, over Bob Sikes Bridge to Santa Rosa Island, really a glorified sandbar. About ten miles down the beach is Fort Pickens; the U.S.S. Forrestal, moored here, is visible across the Bay. No wonder this was viewed as such a strategic location: between Fort Pickens and the Naval Air Station, one commands the entire entrance to the Bay. It is one of the best natural harbors in the Western Hemisphere. And I can easily see why this; or San Diego; Monterey, California (the Naval Postgraduate School); Cherry Point, North Carolina; Jacksonville, Florida; Camp Lejeune, North Carolina; Camp Pendleton, California; and others where I would soon visit would not be bad places to retire: comfortable living conditions, clean, no crime, nice weather, subsidized golf, dining, and medical care. Not a bad life.

Back at the base, I held my smaller seminar discussions in the late afternoon. They also go well; the "students" are grateful. Later that evening several NWC colleagues, also here to lecture on their specializations, and I go out for drinks at the Oyster Bar. Of course, this is also paid for by the taxpayers. On other occasions, when I similarly came to Pensacola to lecture to reserve Officer groups, I would meet with academic colleagues and friends, Alfred Cruzen and Charles Anderson, who tried to recruit me at UWF. But on this first occasion, the main topic of news and gossip was NWC and the colleagues there. To wit:

- A female medal award winner at NWC turned out to be a stripper. After the commandant had pinned the medal on her, she said she now had an award for him and bared all.
- A military officer at ICAF, our sister college at NDU, turned out to be a Satan-worshipper. How did he get through the screening process? His name was seen published in the *Post* as a member of a Satan cult.
- Our colleague Hugh De Santis failed to show his ID badge to security at the NWC gate and said something insulting (fortunately, not racial) as well. For this he was called before the commandant, balled out, and told he would lose his job if it ever happened again. On another occasion, Hugh was bodily hauled out of NWC by Military Police for too often violating the base speed limit.
- The commandant is portrayed as insecure, not an educator (but thinks he is), with no interest in the *substance* of our courses but a stickler for form. Lecturers who go over the time limit or who displease him—e.g., defense specialist Larry Korb—are not invited back.
- NDU has, among its staff, some real mental cases but who, because of civil service rules, cannot be fired. I closely knew several of these; what to do with them when they don't work and yet can't be fired?

An enjoyable dinner. Both subsidized *and* paid for. Not bad. And interesting revelations that NWC is not quite so staid and uptight as thought.

Maxwell Air Force Base: The Air War College, Montgomery, Alabama

I've been to Maxwell Air Force Base and the Air War College (comparable to the Naval War College at Newport, Rhode Island) there, in Montgomery, Alabama, on several occasions. Sometimes I was able to fly there, using the Air Force's fleet of private jets. As a member of the NWC faculty, I was able to do that if there was another high-ranking Air Force officer or DOD official heading in the same direction at about the same time. I'd go from NWC to Andrews Air Force Base near Washington where Air Force One is hangered, go directly to the waiting plane, and off we'd go. Whatever you think of bankers and General Motors executives and their private jets, I've decided that I like to fly by private jet! They're fast, quick, no waiting, no hassles, no security, no lines, superb service. They wait for *you* and take off when *you're* ready, not the other way around.

But on this occasion we were flying commercial from Washington National into Atlanta and then on to Montgomery. Martin Sheen, the actor, was on board, along with a companion that my mother would describe as a "floozy." I took an instant dislike to Sheen. Not only did he try to order the cabin crew around and get special privileges, but as an actor he seeks to use his celebrity status to advance his often far-out political views. Not content to be a "mere citizen" like the rest of us, he demands special access as a celebrity. But why should we listen to him or Barbara Streisand (did she ever go to college?) on matters of foreign policy or national security? They have no more knowledge than any other citizen, and certainly no special expertise.

Maxwell Air Force Base is another of those bases that looks like a country club. But here, unlike Pensacola, the style and architecture are southern: lots of stucco, palm trees, and Spanish moss. The accents on base are, similarly, mainly southern and the style of life is gracious. There are beautiful homes for the officers; all the amenities and services of a small town; the lawns are immaculately manicured; everything is subsidized; and all this luxury is hidden behind a walled, gated security perimeter. Of course, one also sacrifices, sometimes as in Iraq or Afghanistan, making the ultimate sacrifice, as part of the military career; but I'm always amazed at how cushy, especially in peacetime, the life of an officer can also be. If there are to be savings (the post-Cold War "peace dividend") in the Defense Department budget as was then, in the early '90s, being contemplated, these country-club-like bases would be a good place to start.

At Maxwell I've lectured both to the Air War College students there and to the RCNSC (reserve officers) groups that go there for refresher courses.

This time it was for a RCNSC group. The lecture, on U.S. policy in Latin America, went especially well. The students, about 150, are grateful for my lecture and eager to learn. I try to set forth the concept of a post-Cold War "opportunity-based" foreign policy in Latin America, to take advantage of trade, energy, and political/diplomatic opportunities there; but these students, as well as DOD in general, are so accustomed from the Cold War era to the concept of a threat-based foreign and security policy that it's hard to get them to see the other viewpoint and the advantages it provides. We have a spirited discussion, and the students give a little cheer at the end.

I have good friends at the Air War College, independent of any NWC or DOD connections, and I'm able to meet with and have lunch or dinner with them. Bob Wenzel, a colleague from graduate school days, is professor of national security policy and chairman of the department at AWC. He and I were both students of Frederick Hartmann, author of one of the leading textbooks in the international relations field and, along with Hans Morganthau and Henry Kissinger, one of the foremost "realists" in the IR field.

Bob and I have a lot of news to catch up on: he has been in the Professional Military Education (PME) field longer than I and knows it well. He is able to confirm many of the ideas I've already formed on the basis of my short experience at NWC: that the military does not respect academic expertise, that all the "widgets" (including faculty) need to be interchangeable, that original research has no value, and that military officers can't believe that their training does not give them the same ability to teach at this high level as the civilian faculty. But Bob also tells me the interesting fact that the Air War College, unlike NWC, has a path to full tenure for its civilian faculty. Apparently AWC was able to convince the Defense Department that it would never attract a good civilian faculty to Montgomery unless it was able to award tenure. However, DOD seemingly drew the line there: other war colleges, including NWC, would not be allowed to award tenure. That revelation, if the issue had already been decided, made all my work of that year, as a member of a faculty committee exploring the tenure issue at NWC, seem like a waste of time.

My other friend at the Air War College was Prof. Judith Gentleman. Judy was also a Latin America scholar (Mexico) whose previous position had been as a professor of history at the University of New Hampshire. We had become acquainted through our mutual interest in Latin America and the professional association meetings (the Latin American Studies Association—LASA, the New England Council on Latin American Studies—NECLAS) we both attended. Judy was married but her husband's job was in Birmingham; the Air War College was the only academic position she could find nearby; and they lived halfway between the two cities—"out in the boonies of rural Alabama," as she put it. But Judy was smart, balanced, and level-headed, as well as quite beautiful; she adapted to the Air War College better than I did to NWC, mainly by keeping her mouth shut and not objecting overtly to the war college culture and practices that I found objectionable. Whenever I went to Montgomery, we would try to get together; in private conversations we would share our common consternation with how the PME schools worked and the gaping cultural gaps between the academic and the military worlds.

Bob Wenzel and I had drinks that evening at the Maxwell Officers' Club. The night before a group of us had gone to dinner at the private off-base Cat and Harry's where the meals were $50 per plate (again, all paid for by the taxpayers), but tonight we eat at the OC; a great filet mignon, salad bar, and all the wine you can drink for $10. The next morning I had a full breakfast (three eggs, grits, bacon, ham, juice, fruit, breads, coffee—$3) at the gorgeous Cypress Tree Golf Course on the Maxwell AFB grounds. And all this subsidized, no sales tax, and no incentive whatsoever to save money, let alone show a profit. It's a beautiful course, thirty-six holes, a spectacular cedar wood clubhouse, lavishly maintained, verandas and dining areas overlooking the fairways, putting greens, a driving range, with more maintenance and staff people than golfers. The greens fees, again all subsidized, are in the $3 range for active duty and retired personnel. What a life!

This spectacular network of bases around the country and around the world was built at the height of the Cold War when Defense budgets were high. And when we thought we needed to lavishly subsidize the military career in order to keep good people in the service. But as I sat at the Cypress Tree Course eating *my* subsidized meal, I thought that era, post-Cold War, must surely be over. Without an enemy, how could this pampered lifestyle be justified? Surely someone will call for accountability and savings. And then for the privatization of this whole support system. For even now (1993) it's really a throwback to an earlier time and lifestyle that doesn't exist anymore. But while privatization occurred, the cutbacks never did; and then came 9/11, the War Against Terror, and the wars in Iraq and Afghanistan and, hence, the need to make the military career even more attractive. The network of military bases and the perks that go with them remain as plush as ever.

FOREIGN TRAVEL

While at NWC, I engaged in as much foreign travel as domestic. Reviewing my journal notes and c.v. records, I find that these foreign trips averaged about three per year over a five-year period. Some of these, it must be said, were paid by my think tank, CSIS, or other private sources; but the majority came out of War College or NDU funds. Often the travel was to familiar places—Europe,

Latin America—but it also included new and unfamiliar places: Russia, China, South Korea. Essentially, my travel while at NWC turned me from a regional scholar into a global one. And into a comparative politics specialist on a truly worldwide level and not just an area studies one.

Quite a number of my colleagues at NWC were green with envy that I got to travel so much and to such exotic places. But there was nothing sneaky or below-board about this. Because of my writings, I often received invitations to conferences and places that no one else at NWC did. Moreover, as a result of these invitations, I discovered that there were pools of money at NWC and NDU that could be tapped for these purposes. In addition, I'm a quite creative person in terms of rationalizing, through memos, why NWC should not only give me leave to engage in such travel, but should also pay for it. The long and short of it is that, at least three and sometimes four times a year, I was off to interesting places.

I discern three kinds of trips and sponsorships among these many travels. The first are those that came to me as a result of my work and reputation as a scholar, which NWC then paid for but which had little or nothing to do with my official duties at the College. The second are those trips that were part of my official duties at the College: what NWC called euphemistically "Regional Studies." The third category included those trips sponsored by CSIS or that had to do with my own private research, and paid for with outside, non-DOD funding. It all added up to a lot of travel, a lot of what in published form I called "Adventures in Research."[2]

Russia, Spring 1992

In early 1992, still in my first year at the War College, I received a communication from Nizhny Novgorod University in Nizhny Novgorod, Russia. It was from Andrei Makarychev, a young Professor of Political Science at the University. He and a team of his collaborators had been translating my writings on comparative politics and foreign policy and were publishing commentaries on them. Now they wanted me to come to Nizhny and give a series of lectures on these themes. Not only had they done these translations without any royalties or attention to copyright laws but, since their university was without funds, I would have to pay my own way to Russia.

For the first thirty seconds or so, I was appalled by the cheekiness of this request. First of all, you do not borrow or do a translation of someone's work without permission and an honorarium being paid. Second, you do not do any of this in violation of copyright laws. And third, you do not invite someone to a conference and for a series of lectures, and then ask him to pay his own travel besides. I'm enough of an established author that I don't do freebies like that.

But after I'd thought it over for half a minute, I thought, "Why not?" I'd never been to Russia before. It was an exciting time to go: the Soviet Union had just been overthrown a few months before and Boris Yeltsin was struggling to organize a new economy and a new democracy. Besides, my interlocutor in this communication, Prof. Makarychev, sounded so enthused about my work and so thrilled to be talking to the author, how could I resist? I told him I would see what I could do.

So I went to see my friend (and dean) at NWC, Roy Stafford. I told him of the nice invitation and opportunity I'd received. Stafford recognized this was prestigious for the College as well as for me. And at this time DOD, in its bureaucratic way, even though the USSR was already defunct, still concentrated most of its resources on the Soviet Union, now Russia, and its "threat." So Roy made the plane fare available; later on, when it appeared my Russian hosts would be shorter than promised on hospitality, a generous per diem was authorized as well.

So off I went in April 1992 to Russia. I spent nearly a month there. What an adventure! Since I've already written in detail about this trip in another place,[3] only the highlights will be provided here.

We landed at Moscow's Sheremetyevo Airport. The runway was lined with the stripped-down carcasses of Russian airline planes. They had been cannibalized for parts and whole sections for the planes still flying. The sight did not increase my confidence in Russian airlines—or in Russia.

We were met by Prof. Makarychev and his colleagues from Nizhny and spent several days touring Moscow. My impression: a "Western" and modern city on the surface but very Russian and almost Third World underneath. I visited the foreign ministry, lectured at the Diplomatic Academy (like the State Department's Foreign Service Institute where I'd also taught), and visited the U.S. embassy.

We took an overnight train to Nizhny Novgorod. As soon as you leave Moscow, and especially as you head east, the "real" Russia sets in: vast, primitive, disorganized, loud, pushy, poor, diseased, already paralyzed and breaking down in the wake of communism's fall. Nothing (and no one) seemed to be working. Outside of Moscow and St. Petersburg, Russia is a non-Western and a Third World country. Immediately, I grasped that all the nomenclature we'd been using in my field of Comparative Politics for the last thirty years—dividing the globe into a First World of developed democratic states, a Second World of *developed* communist states, and a Third World of poor countries—was wrong. Russia was *not* developed; it was closer to the Third World.

Nizhny Novgorod is Russia's third largest city, with three million people, but it is almost completely unknown. That is because, as an arms, tanks, submarines, and MIG fighter-plane producing city, it was closed to outsiders. I was one of the first Americans to visit. It had no decent hotels, no FAX machines, no functioning telephone system, no restaurants, no banks, certainly no ATMs. It had a lovely Kremlin (fortress) overlooking the famed Volga River and the vast Russian steppes to the east. Vladivostok on the Pacific coast of Russia was a daunting twelve days driving time away.

I interviewed the arms factory managers. In the aftermath of communism's fall and with the absence of any knowledge of how capitalism worked, they had totally unrealistic ideas about new markets for their products. One wanted to convert his submarines to commercial use and ship Siberian logs under the North Pole to Canada (does Canada need more logs?); another had plans to produce a Russian luxury car (half the cars produced in Nizhny factories did not work) to compete with the German Mercedes. Interviewing these officials, I didn't know whether to laugh or cry.

My Russian hosts were extremely gracious—the other side of this large and complex country. They put me up in a private apartment—by booting out the Russian graduate students who lived there. They stocked the refrigerator and cleaned the apartment. They introduced me to their friends and families, took me on tours of the countryside (Vladimir, Suznal), and brought me to museums and study centers; I saw a *lot* of Nizhny.

Our conferences were held in the Kremlin. I helped the Russians set up new courses, programs, and syllabi on international relations, foreign policy, and comparative politics. They were extremely thankful. A low part of the conference was when local communist officials shoved me aside, took over the microphones, and denounced Western interference in their affairs.

A high point, really a series of high points, came when we visited Arzemas16, the secret Russian nuclear facility down the river from Nizhny. I was the first American ever allowed to visit. To begin, although I have no medical training, it looked to me, shades of Chernobyl, that half the town had leukemia or had been contaminated in other ways. Second, I stayed at a Russian military base: after an afternoon of too much vodka, I actually hit (pure luck) the ducks flying overhead, while the Russian sharpshooters with me (*way* too much vodka) missed. Third, this was the spring of 1992, four years before the Nunn-Lugar bill enabled us to get a handle on "loose" Russian nukes. But between 1992 and 1996, the Russians told me, all kinds of weapons, including nuclear materials, were shipped down the Volga to the Caspian out of sight of U.S. spy satellites, and from there to Iran, Iraq, Pakistan, and who knows where else. Scary!

While Washington and U.S. embassy officials in Moscow were touting democracy and free markets for Russia, I saw precious little of either of these in my travels and interviews in the great Russian interior. Instead, what I found was chaos, breakdown, dysfunction. The state farms were idle and nonproductive; Yeltsin issued a decree calling for the privatization of land, but no one had any money or credit to buy it; democracy had been proclaimed but statues of Lenin and the old communist leadership still dominated in every town; there were no political parties out in the countryside I saw and no civil society. The government had no reach outside of Moscow; meanwhile, farm machinery sat rusting in the fields; the common people were waiting for direction but no one provided it.

When I went back to Moscow, I reported all this to Ambassador Bob Strauss (with whom I'd worked on the Kissinger Commission on Central America) and the U.S. Embassy. But theirs was an ideological focus—democracy and free markets—and the fact there were none of these in the vast Russian interior was not a message they wanted to hear. It's always too bad when U.S. foreign policy is designed to play to ideology and domestic public opinion this way rather than dealing with the real world. For that short-term strategy is precisely what produces backlashes in the long term: Putin, authoritarianism, and a reversion to statism and top-down rule. Too bad; they should have listened!

A few years later in 1997, we went back to Russia and Nizhny Novgorod. Constrained by time and the pressure of scheduled formal lectures, our visit on this occasion was briefer and we saw less of the countryside. But the overall impression of Russia was not much different from on the earlier visit: a big and disorganized country, with weak infrastructure and the absence of much civil society, with immense resources but ineffective in exploiting them, weak political parties and legislature, a sense of frustrated nationalism and resentment toward the outside world, a yearning for order and religion. These attitudes are not much different from those found in other large but frustrated Third World nations: Argentina, Brazil, India, Indonesia, Mexico, others. Moreover, and this was already apparent on our second visit, in the absence of strong political parties, civil society, and a functioning government, the desire for strong, nationalistic authoritative, if not authoritarian, rule was already powerful. This was the same kind of public sentiment that brought Hugo Chávez to power in Venezuela; now it would bring Vladimir Putin to preeminence in Russia.

South America, 1992

Each spring, the National War College sends all of its 180-odd students and most of its faculty and administrators on some twenty-to-twenty-five, twelve-day study abroad ("Regional Studies," we call them) trips to the far-flung corners of the earth. The students choose the trip they want to participate in and do most of the arranging; usually, two faculty members accompany each student group. These are not larks or junkets but fact-finding trips designed to educate the students on all the regions of the world. Though there's considerable work involved, these trips are a lot of fun as well; everyone—faculty and students alike—almost always has a good time. In the spring of 1992, since that is the area on which I teach, I was "tasked" to lead the trip to South America.

My group consisted of eight students and two faculty members. We had seven military officers and two civilians—one from State and one from DIA—plus me as the head of the delegation. That role involved a lot of arranging and protocol, which I was not well trained in. It was my job to thank all the groups and organizations with which we met and to present and receive gifts. Sometimes this got awkward with our (mostly) military hosts, since I'm both a civilian and a wise-cracking professor. In addition, as a professional in this field, I often knew more about the politics and history of the countries we visited (Argentina, Brazil, Venezuela) than did our hosts. But we brought it off; as for the protocol, let's just say that I grew in the job.

In Argentina we first had briefings at the U.S. Embassy and met with the Embassy team. Terry Todman, the ambassador, was a friend of mine going back to the Carter era, and so I met privately with him. Our group then had our picture taken with the ambassador; that provided an initial source of friction since the Embassy then insisted we *pay* for the picture. Cheapo!

We had meetings and briefings at the Argentine Defense Ministry, the Argentine Superior War College (like our NWC), the Cavalry Regiment, the National War College, and the Foreign Ministry. Our Argentine hosts were consistently gracious and hospitable, treating us to sumptuous lunches and dinners; I wish I could say the same for our American hosts. The visit to the Foreign Ministry and our briefing by Foreign Minister Torcuato di Tella were arranged through my own contacts in Argentina; the U.S. Embassy did not want us to go there and sought to put up roadblocks. Meantime, in between all these meetings, we functioned as tourists and saw a lot of Buenos Aires: the Casa Rosada, the Congress, the Opera House, Recoleta, tango shows. Of course, I'd been there many times before, but none of the other members of our group had. And, since it's Argentina, we ate particularly well and drank lots of great wine.

It was interesting to me how many little and big frictions emerge in a trip like this. Not within our group—we all got along extremely well—and certainly not with our Argentine hosts, but mainly involving us, the U.S. Embassy, and the military attachés who are our official hosts. First, there was the flap over the picture with the ambassador (we paid, but only to avoid trouble); then, there was the fact the Embassy didn't want us to visit the Foreign Ministry for fear our military guys weren't smart or diplomatic enough.

Even I got angry at this one: who are these State Department elitists to be so condescending toward our War College guys? There were similar disputes as to how high-level our meetings in the Defense Ministry would be and with whom we would meet. The high (or low!) point came on our last night in Buenos Aires when the U.S. defense attachés insisted on *charging us* for the outdoor barbecue they hosted in their private homes. And I had naively thought that was part of their job and that they were happy to see us!

The next day we flew up to Rio de Janeiro. We had meetings and a sumptuous lunch at Brazil's famous Superior War College (Escola Superior da Guerra), with its own private beach and gorgeous views out toward Guanatara Bay. I shocked the Brazilians by speaking to them in Portuguese (from visiting American delegations, they usually get only Spanish) and knowing more about their country (my wife is Brazilian and I've lived in and done research in Brazil) than they did. The commandant of the ESG was a strange bird who had his meals prepared separately from those of the other officers and whose chair was formally placed so far away from ours that (including mine, as head of our delegation) it was impossible to engagé him in conversation. I'm sure that was the purpose of the spacing; maybe I'd not yet learned as much protocol as I'd thought.

I'd also made arrangements through my own contacts to have us meet with the Strategic Studies Center at the Federal University Fluminence in Niterói. This was the only university-based center in all of Brazil devoted to strategic studies; I (mistakenly) thought it would be good for our guys to be briefed there. But the Center had only two active researchers, both were on the Marxist left; neither had ever had any government experience; they knew little except theoretically about strategic studies; and they were strongly anti-American. My fault; I should have known better. It was a complete waste of time, except that crossing over to Niterói we got to see parts of Rio I'd not seen before.

We spent the weekend in Rio doing touristy things: the Corcovado, the cable car to Sugarloaf Mountain, the art museum, the beaches, the Barra da Tijuca (where we'd once lived in Brazil when it was still nice), the Jockey Club. Our hotel was right on Copacabana beach and our guys hung out at the Banana Café where Rio's most beautiful women also partied. Our group was bonding internally by this time with me as their leader and we had a great time, not only with the girls but also with Rio's famed transvestites. Thank goodness ours was a mature and a sensible group. We had spirited debates as to whether Argentine or Brazilian women were more beautiful.

On Sunday afternoon we flew up to Brasilia. We toured the city, its surrounding lakes and residential areas, and the recreational clubs that make life in this isolated capital bearable. The next day we had briefings at the U.S. Embassy and at the Brazilian Defense Ministry.

I was fascinated, again, by the tensions that seemed to swirl up about these meetings and our visit. First, we were snubbed by Ambassador Richard Melton, an acquaintance of mine and a friend and neighbor of one of our group, who for unknown reasons refused to meet with us. Second, our group complained that the U.S. attachés posted to the Embassy had not done adequate preparation in terms of arranging meetings and briefings for us. Third, our military officers, more sensitive to issues of rank than I, said that their meetings with Brazilian counterparts were too low-level, that the Brazilians had shoved them off onto a group of junior officers who were too low-ranking and didn't know much. It turned out that some of our guys (not I) even complained to the Pentagon about our treatment in Brazil, and soon the cables were going back and forth.

The last and perhaps most interesting stop on our trip was Caracas, Venezuela. Venezuela was (1992) extremely tense at this time: several coup attempts were in the wind; Hugo Chávez had just been jailed for launching one of them; there were questions about the loyalty of the officer corps and of specific military units; and Venezuelan democracy was hanging in the balance. And there we were—a group of U.S. military, intelligence, and diplomatic personnel landing in the country in the midst of this crisis for, from the Venezuelan viewpoint, who-knows-what purpose.

As usual, our initial briefings were at the U.S. Embassy. Unlike our previous two ambassadorial stops, Ambassador Mike Skol, another Washington friend and colleague, was both extremely hospitable and very forthcoming in the meeting we had with him. We also visited the military headquarters at Fort Tiuna in the heart of Caracas, the Ministry of Defense, the headquarters of the National Guard, the Venezuelan National Security Council, the foreign affairs Committee of the Congress, and CAVIM, the Venezuelan arms manufacturing industry. PDVSA, the famed Venezuelan state oil company, treated us to a super lunch; in my free time I went out with some of my former graduate students who lived in Venezuela. In all these visits, one could sense the high political tension that existed.

Fort Tiuna was a hotbed of the military and military-cum-civilian conspiracies; our schedule took us back there several times. On one of these occasions the Venezuelans put on an arms display for us. As head of the delegation, I was given the seat of honor, one that had previously been occupied by dictator Marcos Pérez-Jiménez. Then the tanks and armed personnel carriers all pulled up with their heavy cannons all aimed at my chair. I have to confess, with all the tension and questionable loyalties in the armed forces, that at that moment I had only one thought: Anwar Sadat, the Egyptian president who had been killed by his own armed forces while sitting in a reviewing stand. Fortunately, that did not happen to me, but the thought had crossed my mind.

The U.S. military attachés were also worried: the last thing they needed was for our visiting delegation to get in the way or be caught in a cross fire. So the next day they hustled us out of the capital and out of harm's way to Colonia Tovar, a nineteenth-century German community in rural Venezuela. And even as we left the country the following day, there were reports of unexplained troop movements on the radio; the road from Caracas down to the airport was lined with troops; and the airport itself was under tight security. It was not long afterwards that Venezuelan democracy, already wobbly, began to unravel and Hugo Chávez came forward as the new socialist man-on-horseback.

Central America, 1993

The next spring we went on a similar two-week junket ("Regional Studies") but this time through Central America.[4] I was again the head of the delegation; we had seven military officers and one civilian CIA official as part of our group. The purpose was fact-finding on Central America after the signing of the peace accords that ended the decades-long civil wars in the region. We visited Panama, El Salvador, Costa Rica, and Mexico.

Once more, our trip was dogged by controversy even before it began. First, the military attachés in Mexico asked our students to cancel our visit, arguing that Mexico was "too volatile" and they "too busy" for us to come there at this time. They were arrogant about it besides. I had to inform them, using my personal connections, that U.S. Ambassador to Mexico, John Negroponte, was a friend of mine; suddenly the arrangements for the trip began to go smoothly.

Then it turned out that Southcom, the U.S. Military Command for the Southern Hemisphere, insisted that all DOD delegations entering Latin America pass muster with Southcom first. We would have to go initially to Southcom headquarters then in the Panama Canal Zone, receive briefings from Southcom officials, and receive their approval before we could enter the "theater"—i.e., Central America. Thinking like the civilian academic that I am, I resented Southcom's interference, its insistence that we follow Southcom orders, that we get briefed by them first, and that we needed their permission to enter the area. As a result, we had to reroute all our airline tickets and reschedule our entire trip. But I held my temper and resentments in check and it turned out to be worth it; the Panama visit turned out to be among the most interesting of our trip.

We arrived in Panama on May Day, 1993. It was an exciting time to be there. Only two years earlier, the U.S. had invaded Panama, captured Dictator Manuel Noriega, and occupied the country. While some Panamanians applauded Noriega's arrest, others were seething at the U.S. intervention. The new Panamanian President, Guillermo Endara, was corpulent, corrupt, and enveloped in a messy family situation involving his wife and his mistress. Meanwhile, a new election campaign was underway in which the candidates included left-wing singer Rubén Blades, and Embassy favorite Billy Ford, famous for the bloody white shirt he'd worn on TV as part of the earlier anti-Noriega protests (actually it was somebody else's, not his own blood, but it made for "good television"). At the same time, the U.S. was preparing to turn over the Canal Zone and all its facilities to the Panamanians, and that was generating a whole series of controversies.

What fun! At least from my point of view, but not Southcom's. With the Embassy and Southcom supplying the vans, we toured all around Panama, on both sides of the Canal and from Balboa on the Pacific side to Colón on the Caribbean. It's a small country; you can see a lot in a short time. We interviewed leading politicians and government officials from all the political parties. I'd been in Panama several times before but I'd never had access to the military bases (we had fourteen of them in Panama) before; they looked like tropical versions of the country club-like bases I'd seen earlier in the U.S., complete with plush housing, movie theaters, shopping centers, etc. And to think we were about to turn all these multi-billion-dollar facilities over to corrupt and irresponsible Panamanian politicians.

The Panama part of our trip had three big highlights. The first was visiting the Canal, under military escort, which gave us wonderful access. The Canal, with its all-gravity generated locks and water supplies, is an engineering marvel. We were able to visit the locks and guardhouse when a big cruise ship was just passing through. The ships, designed specifically to navigate the Canal, came so close to the shore that we could literally shake hands and carry on a conversation with the passengers as they passed by. I still have in my closet a "Panama Canal Pilot" cap that I wear on special occasions.

The second high point was a visit to Quarry Heights, which was Southcom headquarters in Panama. With its plush housing, commanding views, and lush living conditions, The Heights was an almost perfect incarnation of U.S. imperialism in Latin America and the world. I'd seen this style of architecture before in the Philippines, the Dominican Republic, and on United Fruit Company lands in Central America: colonial design, broad verandas, spacious lawns, gorgeous bougainvillea, local peasants and maids to do the hard labor. But Quarry Heights is imperialism's most blatant expression anywhere in the world; it was almost enough to make you think that maybe Marx had it right after all. Seeing Quarry heights and the Canal Zone—a little slice of the U.S. within Panamanian territory—also helps you understand the reluctance of the "Zonians" (those who lived there on a permanent basis) to give up this wealthy enclave to the Panamanian government that would doubtless run it into the ground and use it for patronage purposes.

The third high point, at least from my point of view, was the fight we witnessed between the U.S. embassy and Southcom over the fate of these bases. The Embassy, under the tough and acerbic Ambassador Deane Hinton, wanted for diplomatic reasons to turn over the entire Panama Canal Zone and its bases as quickly as possible. Southcom, then under equally tough General George Joulwan, wanted to delay the give-away as much as possible and was even then negotiating with the Panamanian government to retain four of the fourteen bases. It was an epic, intra-bureaucratic battle that we were able to observe from the inside in our visits to both the Embassy and Southcom, eventually won by the Embassy with the aid of the Clinton White House.

From Panama, having been briefed on all these conflicting crosscurrents, we flew up the Central American isthmus to El Salvador. Here, too, I can provide only the highlights of what was an incredible visit. As usual, we stayed in the best hotels, ate wonderful food, and interviewed a variety of Salvadoran party and government officials. We were briefed by all the Embassy mission chiefs; the chargé, Pete Romero, was a former student of mine at the Foreign Service Institute, which did wonders for our group's access.

During our visit the Embassy made its fleet of Huey and Cobra helicopters available to us. What a way to see the country! In my youth I'd toured all over El Salvador in "*busitos*" with no springs; helicopters are definitely faster and better. We went the length and breadth of the country visiting Salvadoran military bases and ending up for lunch on the Gulf of Forseca. From there we went high into the mountains along the Salvador-Honduras border to visit the guerrilla headquarters at Perquín. I knew the El Salvador peace accords had recently been signed, which called for the guerrillas to turn in their guns, but our choppers had armed guards riding on the skids with Ak-47s, and we learned that none of the guerrilla guns had yet been turned in. Riding up those mountain passes we were sitting ducks.

We landed on the soccer field in Perquín in a cloud of dust. We were met by a thousand little kids and Ms. Linda Miller, an American hippie (sandals, hair to the waist, no bra) who'd come to El Salvador to join the revolution and fell in love with both it and a revolutionary. Now at age fifty, she needed a life. She'd founded a museum in the town dedicated to the "heroes of the revolution" and whose most prominent display was the U.S. helicopter the Marxist guerrillas had shot down. Now she and her photographer were angling to get a picture of our colonels in front of the downed chopper that she could use for left-wing propaganda purposes. I saw what was happening and managed to evade the photographer; I'm not sure the rest of our group did. So probably out in the Marxist press somewhere there's a picture of

National War College colonels standing in front of a downed U.S. helicopter with the caption undoubtedly (and falsely!) reading something like "U.S. colonels visit guerrilla headquarters, express solidarity with Salvadoran revolutionaries, express regrets at U.S. policy in El Salvador."

From El Salvador we flew back down the Isthmus to Costa Rica. Once again, we can mention only some of the highlights. I'd been in Costa Rica many times before but to our colonels Costa Rica was so much different from what they'd seen in Central America before: green, prosperous, middle-class, literate, democratic, stable. Why couldn't the rest of Central America be more like Costa Rica—the same question I'd heard earlier from congressmen when I'd testified before their committees. The answer, of course, is that Costa Rica has a different culture, history, and sociology than these other countries, and you can't just plunk one country's culture and institutions down in another country's territory and expect it to "take."

Second, in the U.S. Embassy at this time there was no ambassador. But the Embassy staff had heard that a number of Washington political types were campaigning for the job. I knew all the personalities involved, and passed what I knew on to the Embassy. Pretty soon the calls between San José and Washington were flying back and forth. The long and short of it is that none of the self-promoting candidates got the ambassadorship.

A third issue was the quality of the U.S. military attachés in Costa Rica. Since Costa Rica had no army of its own, this was a low-prestige post for a U.S. attaché, and the quality of the people sent there was very low. I had nothing to do with this one but our military guys did, incensed at the incompetence of the U.S. attachés and their attitude. When we got back to Washington, they sent a scathing letter to the Pentagon; it wasn't long before the U.S. military personnel in Costa Rica were replaced. It was interesting to observe this process as it worked its way out.

We took side trips to both Punta Arenas and its beach on the Pacific side and Puerto Limon on the Caribbean. In Puerto Limon we boarded a police launch and went out to the offshore island where Columbus landed on his voyage to this area. We then went out into the open sea, north past the port of Moén and then upriver into the Costa Rican rain forest. We were ostensibly looking for druggies (and I have pictures of me on the prow of the boat looking Rambo-like with an AK-47) but in reality what I mainly saw were spectacularly colored birds, monkeys, even caimans.

Far upstream we came to a fisherman's shack whose owner doubled as a drugs informer for the police. He must have been used to gullible American tourists descending upon him because he told us if we just went upstream a few more miles we'd discover—are you ready for this?—both ex-CIA director Bill Casey and former president Richard Nixon still living and with native

Costa Rican girlfriends. We all got a good laugh out of that (when your boat subsequently got stuck while looking for Casey and Nixon or you sheared an outboard pin, he'd rescue you—for a fee of $100); meanwhile, the "fisherman" supplied us with a case of very welcome and delicious Costa Rican beer.

Our last stop, back up the isthmus, was Mexico. On this trip, we'd also stopped briefly, if only at the airport for a quick briefing, in Nicaragua, Honduras, and Guatemala so we'd really seen a lot. Mexico was our most difficult stop, mainly because the head of the military group, a brigadier general, didn't want us to come there at all and continued to put up- roadblocks to our visit. But we persevered; I intervened on our behalf with my friend, Ambassador Negraponte, and we ended up seeing and doing a lot despite the attachés' hostility.

Mexico is a highly nationalistic country, particularly with regard to the U.S., and it puts severe restrictions in the way of contact between the U.S. and the Mexican militaries. Nevertheless, and no thanks to the U.S. attachés, we had lengthy visits at the Defense Ministry, the Mexican National Defense College (our counterpart), and the Foreign Ministry—again, thanks to my contacts there. Our Mexican military colleagues were, perhaps surprisingly after the earlier comments, remarkably hospitable, friendly, and cooperative. We had a great time and learned a lot.

But the U.S. attachés continued to try to sabotage our trip and especially our military and Foreign Ministry contacts. One day they shuffled us off to Cuernavaca and the Nissan automobile factory there as a way of getting us out of Mexico City and, presumably, out of harm's way. But we also learned a lot there and had interesting conversations about drug trafficking, immigration issues, and U.S.-Mexican relations. Another day they sent us up to Teotihuacan to climb the pyramids, a junket that with my interest in comparative civilizations and their development I found fascinating, even if it had little contemporary military relevance. Plus, we all fell in love with our Mexican guide, Guadalupe (what else?) Escalante, who proved to be not only very *simpática* but also extremely well-informed on all aspects of Mexican politics and international affairs. So much for the attachés' efforts to keep us insulated from Mexico and Mexicans! When we got back to Washington, our guys also sent in a scathing report on these attachés.

Other Trips

While at the War College, I also engaged in other foreign travel. In five years at the College, I must have undertaken about fifteen foreign junkets, or about three per year. Some of these were paid for by DOD; some were partly sponsored by the Defense Department; still others were sponsored by my think tank, CSIS; and, finally, other sponsoring agencies (USIA, State Department) paid for some of these trips. The War College and DOD were very generous with their funds and in allowing me the time and opportunities to do all this travel away from my official duties. I am grateful to them.

Even when traveling with non-DOD funds, I had my NWC calling cards and, depending on the circumstances, often identified myself as with the War College. That got me into NATO headquarters, European defense ministries, and U.S. military bases abroad to which I would not otherwise, as a civilian, had access. So I'm grateful to NWC in this way as well, because my affiliation there got me into places, provided me with information, and added enormously to my storehouse of lore that I would use in later research projects and writing.

Because these were often private or semi-private trips, and because I've written about these travels at length elsewhere,[5] I intend to be very brief here and only list the travel undertaken and its main purposes.

In 1994, 1995, and 1996, I was in the Dominican Republic, a country about which I'd written my first books, as an election observer (1994 and 1995) and for a conference (1995) on emerging leadership in the Caribbean region. These trips were sponsored by my think tank, CSIS, though NWC gave me released time.

In 1995 I was in Jamaica for another CSIS-sponsored conference on the Caribbean.

In 1993 and 1996 (twice) I was in Portugal, Spain, and Western Europe working on a research project and as an election observer; NWC provided the funding.

In 1994 I was in Asia (Japan, China, South Korea) to give lectures and involved in another research project; funding was provided by NWC, the Korean and Chinese governments, and Japanese business and university groups.

In 1995 I was again in Venezuela working on a survey of public opinion sponsored by the U.S. Embassy, the COPEI (Christian Democratic) political party, and the International Republican Institute. Our survey showed that Venezuelans wanted what they shortly got in Hugo Chávez, a leader who was strong, nationalistic, authoritative, nonpartisan, socially reformist.

In 1996 I was in Japan to give a series of lectures; NWC covered the travel expenses.

In 1997 I was again in Russia; this time USIA paid for the travel expenses.

These trips were in addition to the other trips previously recounted and to the "Regional Studies" that we undertook yearly. I profited greatly, as recounted above, from the Regional Studies and am grateful to NWC for sponsoring or partially sponsoring a number of the other trips.

It remains a bit of a mystery to me how I was able to do all this travel and why my NWC colleagues did not take advantage of the same opportunities. Here's how it worked for me. I only discovered this opportunity by accident in my first year at NWC when I applied for funds for my first trip to Russia.

It seems that every year NWC had surplus funds in its travel budget. This was money left over from the College-wide Regional Studies trips. But this fact was not advertised nor, unlike in a civilian university, was there a regular process for applying for research and travel funds. At the same time, at the end of the budget year NWC would be loathe to report a surplus of travel funds or the next year its travel budget would be cut. The secret was timing: to get in my request after Regional Studies (the big travel item) had been budgeted but before the end of the fiscal year.

So every year when classes were ending in late April, I'd go to see XO Colonel Kirk Lewis who controlled these funds. Since NWC was eager to spend its surplus before the fiscal year expired, it didn't take much of a justification—lecturing abroad, research, planning for a future course—for Lewis to OK this travel. I was told "not to go over $4,500," which was a nice stipend for travel and per diem in those days over the (usually) two-week period that I was away. The NWC Foundation also provided funds for this travel. NWC thus provided me with some wonderful opportunities to travel, do research, see places (China, Russia, South Korea) I'd never been to before, and vastly expand my knowledge of foreign and international issues.

Between Regional Studies for NWC, my research travel sponsored by CSIS, and these additional private trips paid for by the War College, I really saw a lot of the world—Europe, Russia, Asia, Latin America—during the early-to-mid-1990s. In the following years I would have the opportunity to explore in depth those parts of the world I'd not visited while at NWC: Eastern Europe, South and Southeast Asia, the Middle East, and Sub-Saharan Africa. This travel and the research associated with it, on civil society, development, and comparative democratization, completed my journey from an area specialist (Iberia and Latin America) to a genuinely global comparativist and foreign policy generalist. I will always be grateful to NWC for the opportunities it provided.

NOTES

1. See, for example, Howard J. Wiarda, *Introduction to Comparative Politics* (Belmont, CA: Wadsworth, 1993); Wiarda (ed.), *U.S. Foreign and Strategic Policy in the Post-Cold War Era* (New York: Greenwood Press, 1995); Wiarda, *Non-Western Theories of Development* (Fort Worth: Harcourt Brace, 1998); Wiarda, *Comparative*

Democracy and Democratization (Fort Worth: Harcourt Brace, 2001); Wiarda, *Political Development in Emerging Nations* (Belmont, CA: Wadsworth/Thomson, 2004); Wiarda (ed.), *Comparative Politics* (London: Routledge, 2005), 6 vols.

2. Accounts of the trips undertaken while at NWC are in Howard J. Wiarda, *Adventures in Research,* Vol. III: *A Global Traveler* (Lincoln, NE: iUniverse, 2006).

3. *Adventures in Research,* Chapter 19.

4. *Adventures in Research,* Chapter 21.

5. Adventures in Research, Vol. III.

Chapter 6

The Military Mind

Is there a military mind? Is there something in the military institution that helps create a military mindset, a way of thinking, a political culture that is different from that prevailing in the civilian culture? And most particularly, that is different from the academic culture?

An alternative way of formulating this question is to ask if the military career of a professional soldier attracts certain types of persons to the armed forces? Which, therefore, by self-selection, reinforces the notion of a special military mindset. Either way, the outcome is the same: either the military institution itself shapes a particular kind of mindset or it attracts a certain kind of personality. In both cases, the result is the same: a military institution that is separate from, thinks differently than, and has distinct core values as compared with civilian society.

It is one thing to perceive oneself, or to be perceived, as having a different way of thinking than the rest of society. But now add in a new ingredient: not just the rest of society but a particular segment of that society, the academic community or professoriate. The professoriate is, in parallel fashion to the military, an institution with a particular way of thinking, its own mindset, a political culture all its own. More than that, the academic political culture of the professoriate—very liberal, open-minded, critical, free-wheeling—would seem to be on most issues diametrically opposed to that of the professional military.

And that is the subject of this chapter and the following: the inevitable clash, certain to occur, when these two opposed cultures are brought together in a single institution such as the War College. The sparks between these two professions and cultures are certain to fly because they start from such diametrically opposed assumptions. There is bound to be a clash in any institution, such as the U.S. Department of Defense, where two cultures, the military and the civilian, are brought together in close proximity and expected to work together closely to solve common problems. But the problem is compounded when it is not just civilians who are brought into a military institution, but civilian *academics*, the *professoriate*, whose values and mindset are so at variance with the military mind.

It is our thesis in these chapters that this kind of clash is bound to occur, and likely to lead to all kinds of conflict, especially in the academic institutions, the PME (Professional Military Education) schools, run by the armed forces. Or perhaps in some of the military think tanks or the policy shops, such as the RAND Corporation or the Center for Naval Analysis (CNA), where civilian academics are heavily represented, or perhaps in the Office of the Secretary of Defense (OSD). These kinds of clashes will be endemic wherever these two cultures, the military and the academic, are brought together in such close proximity. The different PME schools will have different ways of meeting or resolving the inevitable conflicts that arise, but of the *fact* of their existence there can be no doubt.

There is another "clash of cultures" present at the War College, but we cannot give that much attention here. And that is, within the student body, between the military officers and the (mainly) State Department civilian students. Recall that each NWC class is three-quarters military and one-quarter civilian. Recall also that at the NDU level there is a vice-president who is a State Department senior ambassador; at NWC there is also a ranking ambassador who serves as the International Affairs Adviser and, in my time there, was "dual hatted" as head of the Department of National Security Affairs. The role of these two officials is to advise the NDU president and the NWC commandant, respectively, on State Department concerns, and to smooth any ruffles that may come up between the military students and the foreign service officers. And there are *many* of these, stemming mainly from the fact NWC is basically a military post, an Army base, and has a curriculum oriented much more toward military officers than toward FSOs. And because the military officers and the FSOs come from such different backgrounds, perform differently in the classroom, come from such different cultures, have different politics, and see the world so differently. During my time at NWC I was fully aware of these differences but, since I did not have to deal with these issues directly at an administrative level, I only touch upon them here but do not go into any detail.

In this and the following chapter we focus on the clash of cultures that is present at the apex of the PME system, the National War College in Washington, D.C. In this chapter we look at the military side of this equation; in the next, on the academic side and the viewpoint of the professoriate.

While the NWC provides the main subject for our examination, we also draw upon the experiences of other PME schools that have arrived at different ways of resolving, or living with, this clash of cultures.[1]

THE MILITARY MIND

Is there a "military mind"? Is there something about military training, the military career, the military institution and profession that creates a special and unique kind of person? Conversely, does the military, with its clear authority patterns, hierarchy, and discipline, attract a certain kind of person to its ranks? Whose ideas and basic values are, thence, reinforced by the military institution? Actually, the answer to all of these questions is a clear "Yes": there is a set of beliefs that we can identify as part of the military mind; there is much in the military institution and profession that creates a unique military person; and there is a certain type of person that is attracted to the military career, whose value system is then reinforced by being *in* the military.

It is not our purpose here to present new or original evidence about the military value system. That has already been done by a large number of scholars.[2] Rather, our purpose here is to show how these military values and ways of thinking and behaving manifest themselves in an institution like the National War College. And how these, in turn, clash with the civilian-academic culture at the College. In the process we may say something original about the military mind, civil-military relations, or the "two cultures," but our actual purposes are more modest than that.

I am using the term "two cultures" in much the same way that British physicist, civil servant, and novelist C. P. Snow used it fifty years ago. It was 1959 when Snow delivered his famous lecture, later expanded into a book,[3] lamenting what he saw as a dangerous schism between the world of science and the world of culture and literary life. Snow saw these as two quite distinct "cultures," with different adherents, belief systems, and ways of behaving. The two cultures had grown so far apart that they not only could not communicate; they also showed signs of dividing into mutually hostile camps.

That is how I see the "two cultures" operating at the National War College—and in the other DOD-run or PME schools. Except that the two cultures I study are not the humanistic and the scientific cultures examined by Snow but the military culture, on the one hand, and the civilian-academic, on the other. Like Snow, I see these two cultures as dangerously drifting farther apart and diametrically opposed. Both cultures have their own practices, belief systems, and ways of thinking. It is hard for them to communicate and they frequently divide into hostile camps. That, of course, diminishes their effectiveness. So the question explored here is whether these two cultures can *ever* be reconciled, whether they can work together, or if they are so far apart as to constitute permanently hostile armed camps.

It should not be surprising that we can identify something that we can call the "military political culture" or the "military mind." After all, people who come from similar backgrounds, act the same way over a long period of time, interact mainly with each other, are socialized in the same training programs, are part of the same institution and profession, and identify closely with each others' values, tend to develop distinctive and persistent habits of thought. All professions function this way, whether we are talking of lawyers, doctors, plumbers, or, in this case study, military officers and civilian academics. All of us tend over time to absorb and inculcate the values and belief systems of the profession or institution of which we are a part. In the case of the military, the question is whether, given the special nature of that institution, it holds to the values of the profession even more strongly than in the case of other professions. And if those can *ever* be accommodated to the values of the professoriate.

At the same time, let us quickly introduce some caveats by way of qualification. First, the "military mind" is an "ideal type" in the Weberian sense; it is an abstraction that is useful for analytical purposes that need not correspond exactly to reality in all cases. Second and related, it should be clear that no one individual or group will adhere exactly to all the distinct facets of the military ethic, since no one is ever driven exclusively by one's military or any other profession. Other values and beliefs also come into play. Nevertheless, there *is* something, a set of beliefs, values, and ways of behaving, that we can identify as part of the military mind.

No better source exists for an exploration of the military mind than Samuel P. Huntington's classic, *The Soldier and the State.*[4] Although modified over time, Huntington's analysis of how the military institution thinks and behaves is still right on target.

Huntington's research leads him to conclude that the military mind (a term he uses) is disciplined, orderly, logical, and "scientific," in the sense of looking for evidence and hard facts for its assessments. He contrasts these virtues with others that he finds in the academic community, which he sees as tolerant, intuitive, open-minded, and often deductive. He goes on to describe other traits of the military as authoritarian, top-down, bellicose, and often rigid. The military, he says, sees violence as rooted in the permanent biological (aggressive) and psychological (selfish) nature of mankind. Therefore, the military must organize to combat these human tendencies on the basis of organization, discipline, and strong leadership. Consequently, he sees loyalty and obedience as the highest military virtues. There must be a strict chain-of-command;

orders must be carried out; in the military institution, there is little room for dissent. From the military's point of view, *everything*, including dissent, individual rights, and other possible objectives, must be subordinated to the success of the mission.

Huntington goes on to explore the tension between the military institution and its norms, *and* the values of an open and democratic society. He does not resolve this issue but emphasizes the tensions that may exist. Similarly, with individual conscience: there will at times be a conflict between military orders and personal morality—another unresolved issue. In parallel fashion, Huntington emphasizes that there will always be a divide between the corporate or institutional requirements of the armed forces and the individual civil or political rights (freedom of speech, assembly, press, religion, etc.) of its members.

Huntington sums up his chapter on "The Military Mind" by characterizing it as an institution based on "conservative realism." The good of the whole or the mission must come before the interest of any one individual or subgroup. Order, hierarchy, and division of function are the main organizational principles. The military stresses continuity and the value of history. The nation-state is the highest form of political organization; loyalty to the nation and patriotism come above all else. The military ethic, Huntington concludes, is thus pessimistic, corporatist or collectivist (in its loyalty to the institution), historically inclined, (hard) power-oriented, nationalistic, militaristic, and instrumentalist.

I found all of these traits to be present in one form or another in the military structure at the National War College. The first question that arises is: Can you run a serious college or university on this basis? And the second question is: Can these traits of the military mind or the military institution be made compatible with the requirements of an increasingly civilian and high-powered academic faculty? In this chapter we focus on the military side of these issues; in Chapter 7 we consider the issues of most interest to the civilian faculty.

THE COMMAND SYSTEM

The National War College and its parent organization, the National Defense University, are organized and run on a top-down, military-command system. First, NDU is a creation of the Joint Chiefs, not of the more civilian-dominated Office of the Secretary of Defense. Second, the president of NDU is *always* a three-star general or its equivalent; the commandant of NWC is *always* a two-star. At a third level, the two deans at NWC were always military officers at full-colonel rank; fourth, until recently, department heads were always military officers as well.

Although this military command structure remains intact, recently there have been some modifications. At the NDU level there has always been a (relatively powerless) vice president, usually a career ambassador from the State Department about to be put out to pasture; now there is a new civilian provost as well, appointed by the NDU president, whose precise roles and influence have yet to be determined. At NWC there is also a career diplomat who doubled as the head of the National Security Affairs Department, but in my time there that position was always subordinate to that of the commandant; recently the XO or Chief of Staff position has been converted into a civilian position. With only one or two exceptions, the deans have always been career military officers; the same for department heads with the exception of one short-time head, who turned out to be more militaristic than the military.

The conclusion: the military remains in firm control of all NDU and NWC positions, with only a few sops to civilian influence. Even in those few instances where civilian or civilian academics have been appointed to important positions, they are appointed by the military; they serve at the military commander's discretion; and they report to the command structure. I see no evidence of a diminishment of military control at either NDU or NWC levels.

Here are some examples/instances of how this command system works—and its implications:

- When memos about meetings and events go out, they take the form of orders, not requests. Meetings are scheduled for a certain time and you're expected/ordered to be there. It is *not* a system, as in most university contexts, where you consult first about a time convenient to everyone and then schedule the meeting; instead, at NWC meeting times are scheduled in the form of decree laws.
- In my first months at NWC, when the commandant's secretary took a two-week vacation, he simply commandeered our departmental secretary despite the fact she had work to do for us—including a manuscript that she was typing for me. But that's what commandants do; they *command.* No please or by-your-leave.
- One day early in my tenure at NWC, the commandant walked into my seminar at its midpoint and announced that he was sitting in. No request had been made; no by-your-leave; no warning; no purpose stated. He said nothing further and walked out before the end. That's also what commandants do: they check (some would say spy) on their faculty as

if they were drunken sergeants at the gate who had to be watched at all times.

- Decisions are often made by decree law or command. I thought I'd seen the last of that in my studies of Trujillo's Dominican Republic and Salazar's Portugal.[5] Whether its curriculum change, personnel decisions, or administrative matters, *all* decisions are top-down and, often, arbitrary. There's no bottom-up, grass-roots consultative or committee system; instead, we are *told* what to do.
- Faculty contracts at NWC had generally been automatically renewable. But my first commandant at NWC was holding them up as a way of frightening the faculty and putting pressure on them. A number of faculty feared this holdup was preparatory to getting rid of them. Later we heard the commandant had his own list of faculty he wanted hired and was angry when he found out we had three-year contracts and he could not do that.
- I was appalled at NWC over the lack of rules, procedures, and regular processes. Everything was arbitrary, dependent on the commands (whims?) of the commandant. I lobbied at one point for a committee system and a set of rules to govern faculty appointments, reappointments, and governance—only to be told none of this would work or could even be considered because it interfered with the commandant's prerogatives.
- At NWC there is no faculty senate, no faculty union, no organization of any kind to protect faculty interests because that would also interfere with the chain of command and is seen as like an enlisted men's union as in the Netherlands—to be avoided at all costs.
- At one point we began to reorganize Course 4, Geostrategic Context. To help the process I sent a long memo to the commandant, since he would be making the decisions, explaining the logic of the course and what needed to be done. For my troubles, all I received was a reprimand for violating the chain of command by not talking to the dean first, and an *order* that I was to support the decision made whatever it was. So stand up and salute, Private Wiarda!
- Periodically NDU or NWC or both would tighten their procedures. *The command* would speak—whether on increased teaching loads, our obligations to help the military faculty become better teachers, or, as suspected, just to assert its authority. My experience has always been that those who need to assert their authority usually have no or weak basis for their authority in the first place.

There are dozens of other examples and anecdotes that could be listed. The long and short of it is that NDU and NWC, as military bases, operate on the basis of a system of chain-of-command authority. This is emphatically *not* a grass-roots participatory system, a system of governance by committees and consensus and, least of all, a democratic or representative system. Instead, it is a *command* system. One understands why the armed forces must operate on that basis, but one would think that, at the nation's premier war college, some bow in the direction of greater faculty involvement in decision-making would be not only good for the faculty but also beneficial to the institution and its mission.

RANK AND HIERARCHY

NWC is a top-down and authoritative, if not authoritarian, institution in which considerations of rank and command authority are all-important. All decisions are made at the top and conveyed downward through the ranks; there are almost no procedures for grass-roots or bottom-up input into decision-making. How can this system function and be effective in an institution that calls itself a "college" and even a "university"? Once again, our methodology is to present a series of brief anecdotes and cases to illustrate the overall points.

- When I was new at NWC, I naively asked my dean about procedures for determining faculty reappointments, pay increases, and personnel matters. I also asked about procedures for modifying the curriculum, assigning courses, and determining teaching loads. Every college or university I know would have such procedures in place. But the answer I got was, "There are no such procedures." Why? I asked. Because having too many procedures in place would inhibit the commandant's capacity to *command.* That's what commandants do: they *command!* Anything else would dilute their authority and that cannot be permitted.
- At one point, I was scheduled to host, at lunch at NWC, the head of the Spanish national war college, Spain being one of my countries of research specialization. But then the president of NDU determined that I was "too low-ranking" to host a Spanish general. The president took over the hosting role, disinvited me entirely from the lunch, and consigned me to hosting a coffee reception for the Spanish general's aides, at the junior-officer, major and lieutenant, level. Rank is everything!
- Talking with some American junior officers, I discover that the military will pay for a captain or a major to take time out from his career to get a Ph.D. But such well-trained and qualified persons cannot then teach at NWC. Why, I ask my dean? Because our students, at lieutenant-colonel and colonel levels, would outrank them; a colonel will neither listen to nor

have any respect for a major or captain, however better educated he may be. Does this make sense? It may in the service's ranking system, but it doesn't in an academic institution.

- NWC has endless Mickey-Mouse regulations that go far beyond anything at a civilian university. Having to comply with these regulations is the price you pay for working on a military base. Generally, NWC is a pretty relaxed place, but the harsh rules are on the books and they are frequently enforced. These include a dress code, various social obligations, required attendance at military promotion ceremonies, mounds of paperwork, obligatory attendance at all lectures, obligatory committee meetings, and so on. These events are obligatory for civilian as well as military faculty.
- NWC has no procedures for salary increases or merit. I went there with the understanding that I would be eligible for the same yearly step and salary increases as other federal employees. Theoretically, that's possible at NWC, but it never implemented the legislation. Its rationale? Step increases and merit open dissension, envy, and disparities within the "ranks," and that cannot be permitted. Better to treat everyone the same.
- NWC not only has top-down decision-making, but all decisions are made by the military hierarchy. A handful of civilians *may* be consulted (most often they were not, and certainly not systematically), but the decisions themselves are made by the military brass. I doubt if that will change much with the coming of a new civilian provost.
- One day all the CINCs came to NWC for high-level meetings. These are the regional commanders (four-star) who act like imperial viceroys in their respective areas of command. When that happens, security becomes very tight; you can hardly drive on the base; our parking spaces are commandeered and even some of our offices; and the CINCS sweep in with their convoys and large entourages like visiting potentates arriving in a desert kingdom.
- At one of our rare department meetings (probably called for that purpose), we were informed that someone had parked in the commandant's space. The commandant was all excited; this is a no-no; the XO Kirk Lewis was sent to warn the faculty. I didn't do it, though the thought may have occurred to me once or twice; my main thought, however, was that this was petty and sending the XO to "take care of it" represented over-kill.
- One day we received from the commandant a memo on "Faculty Priorities." In the preface he stated he'd conferred with the NDU president, the dean, and department heads; he did not mention consultation with the faculty. His (not the "Faculty's") priorities were:

  a. Teaching

  b. Course design and curriculum development

c. Advising students
d. Additional service to the College and University
e. Scholarship and professional development
f. Outreach

Note that Scholarship is fifth out of six priorities. No mention is made of research and publication—though "scholarship" may encompass that. The real killer came in the last paragraph of the commandant's memo where *only* teaching was emphasized and scholarship was allowed only "when time permitted." Scholarship was seen as "extra-curricular," a spare-time activity.

- The NDU president, an educator and competent in other ways, is seeking to centralize power at the University level. He wants to strip the colleges, including NWC, of their autonomy and concentrate it in his own staff; he also wants all the colleges (ICAF, NWC, etc.) at NDU to follow common procedures and have a common program. He's one of those change-the-whole-system presidents, versus the passive out-to-pasture types whom we much prefer. To us at NWC, that's a disaster; it means more bureaucracy, more dumb rules, more decisions taken farther away from "home," at the College, where the students are. In addition, the president is fascinated by new, computerized teaching machines. That way he could do away with the faculty altogether!
- We are shown a memo that comes down from the Joint Chiefs. The memo makes clear the Chiefs are not interested in the civilian faculty at NWC. The entire emphasis is on strictly *military* affairs, when most of us are specialists in political-military relations, and on *jointness,* in which none of us had much interest or expertise. The Chiefs want to convert NWC into a military strategy school and do away with its foreign policy programs.
- Our commandant is very critical of the curriculum and apparently came to NWC with a plan to get rid of the entire civilian faculty. That was based on his own unhappy experience as a student at the College fifteen years earlier, before Title 10 and the significant improvement in faculty hires. That's the trouble with having no regular procedures at NWC and commandants that are all-powerful. The commandant's views and intentions revive fears that professors at NWC are seen as a "pass-through" or temporary faculty. But in this case, the commandant discovered most faculty contracts had already been signed; he could fire only one person; and he was a very frustrated fellow. Because, before coming to the College, he'd already lined up his own slate of favorite faculty.

- NDU is undertaking major curriculum changes. All this is being decided without the faculty's knowledge or input. The plans are to have us teach more, with more sections, more students, and more responsibilities. The *students* will choose the time and day of the course meetings; faculty may have to teach two or three sections of the same course. These plans echo an earlier comment that the College is run for the students, not the faculty. All these proposed changes reflect a typical chain-of-command decisions model practiced at NDU.
- General Barry McCaffrey was at NWC to give a talk; the College was hosting a coffee for him. I had known Barry for years when he was Southcom commander and had worked closely with him on drug and security issues. But when I showed up at the coffee klatch, I was told it was limited to "senior staff." Well, if senior scholars are not senior staff, I don't know who is. Fortunately, McCaffrey saw me at the door and waved me in, but the next day there were repercussions. Apparently, I was socializing above my rank again.
- Our department head was becoming increasingly intrusive. He had an authoritarian complex that had gone undiscovered prior to his taking on the chairmanship. Now he was constantly looking over our shoulders, sitting in on our classes, and seeking to exercise even the most minute control over our teaching. It was the "drunken sergeant" (the faculty) syndrome again carried to extremes. The head was getting to be too much; we could not function as teachers this way.
- At higher levels, things were no better. Both the dean and the NDU president had advanced academic degrees. But apparently the experience had not rubbed off on them, or it had produced a counter-reaction on their part. Both were huge disappointments, and from people whom, because of their academic training, we expected much more. The NDU president was arbitrary, authoritarian, sought to centralize all power in his own hands, and wanted nothing to do with the faculty. The then-dean was weak, not up to the job, in over his head, didn't want to meet with the faculty, and had no pedagogical experience; he was a chain-of-command type. So, at all levels the system was closed off to advice, consultation, give-and-take. Not a good way to run a college or university.
- I developed a theory about all this. I concluded that the top command at NDU and NWC were chock-full of inferiority complexes. They were unsure of themselves and lacked confidence. They most often lacked the academic credentials of the civilian faculty, lacked knowledge and expertise on the subjects taught, and were not themselves good teachers. So they made up for these faults by being authoritarian, putting the faculty in their place, and by *asserting* their dominance. I think Freud had a term for this—displacement.

TIME

Time at the National War College is not your own. Unlike at a civilian college or university, your time belongs not to you but to the Department of Defense. In my civilian university, in contrast, once I've done my teaching, grading, office hours, and committee assignments, my time is my own to use as I see fit. Of course, at a major research university, it's expected that time will be appropriately utilized for research and writing. But it's *your* time; you decide when, where, and how to use it. If I can let a little secret out of the closet, at a research university like mine where the teaching load is light, I'd estimate that fully 80 percent of my time is mine to use in these ways. Is there any other job or profession in the *world* where so high a percentage of your time is your own? I doubt it.

That is definitely not the case in the Defense Department. The concept of time and who controls it is very different than in a civilian university. This is liable to come as a shock to those like me, used to operating in a civilian University environment and by the relaxed, unregimented rules of academia. For in the Defense Department, obviously derived from a set of rules and expectations enacted for defense personnel, you are "on duty" twenty-four hours a day, seven days a week. The defense of your country, the logic is, including for those of us civilians at NWC, is not a "nine-to-five" job. Rather, it is permanent and all-the-time, at night as well as during the day, 24/7, on weekends as well as during the week. Technically, by military doctrine, there is *never* a minute when you are not "on duty."

The implications of this are immense and subject to a great deal of abuse. It means you can be called and *obliged* to attend an evening event as well as daytime ones. It means you may be called up for "duty" on a weekend as well as during the normal work week. It means you may be obliged to teach extra classes or give special lectures (such as my Reserve Officers National Security lectures) that are not part of your normal teaching schedule or even our employment contract. It means that during the day you are expected to attend promotion ceremonies, going-away events, and other formal ceremonies even if you do not know the person celebrated at all.

Most important for civilian faculty who also have a research and writing agenda, it means you cannot stay home a day or two (or three, as I often did at my civilian university) a week to work on your research there. Instead, you are expected to be *at* NWC and *on duty* every day and all day. Even more troubling, when you're at NWC, you're expected to be working full time on your teaching responsibilities. Research and publication are viewed as purely private or extracurricular activities, which you may indulge in the evening, on your own time (which technically you do not have). One can readily see how this conception of duty time can be used to rule out all independent research and writing, even if that research is directly related to NWC, foreign policy, or national security issues.

Here are some examples of how this policy and understanding of duty time may affect your work:

- As seen in the previous chapter, I was frequently "tasked" to give lectures to RCNSC groups in different parts of the country. I came to enjoy that work, but initially I resented being obliged to do so and to give up time that I'd rather spend on scholarship.
- Ditto for other official lecturing obligations that I was assigned: the Capstone and General Flag Officers' Course, the wives' group, the Inter-American Defense College, at ICAF, and many others. None of these was in my contract, yet since I was "on duty" I was expected to do so.
- The same for all those myriad ceremonial occasions at NWC: we civilian faculty were obliged to attend to help promote College *esprit* even if we did not know the person celebrated.
- With regard to research, secretaries were told not to type my materials because those were purely "private" rather than "official." NWC lacked a third category employed in most civilian colleges called "professional" (professional correspondence, professional phone calls, professional activities in general). Rather, only "official" teaching and course preparation work was permitted.
- On numerous occasions, department heads, deans, and even the commandant walked into my office and were absolutely stunned to see me working at my research. Often these activities were ignored (NWC *does* take some bows in the direction of being a real university), but on several occasions I was told not to engage in research while I was "on duty" (itself a ridiculous concept if applied to academics). Because that time was not my own; it was Defense Department time.

THE MISSION

Almost all organizations nowadays have mission statements. In academia these emanate regularly from deans, provosts, and college presidents. We faculty generally take these statements with several grains of salt. We've heard it all before, often ad nauseum. Our eyes glaze over. We already *know* what our mission is: teaching, research, and public service. All else is usually rhetoric—or, nowadays, fund-raising.

But not so in the Defense Department or the War College. There, *The Mission* is taken very seriously indeed. All else is subsumed to the overriding goal of accomplishing The Mission. The Mission has a hallowed place in military history; in some respects, it is a throwback to earlier campaigns. If The Mission is to "take San Juan hill," or capture Iwo Jima, or hold the Philippines, or take the beaches at Normandy, then you had better do so. And all else can legitimately be sacrificed for the sake of accomplishing that goal. Especially in time of war or national crisis.

But how does that apply to an educational institution like the National War College? And especially in times of peace, such as the mid-1990s when, post-Cold War and pre-9/11, the U.S. had no known enemies and there were few serious threats? The answer—and the possible abuses—is remarkably similar to those discussed under the concept of time. For at the War College our principal mission is the education in national security policy of future flag officers. That is The Mission. And to that overriding purpose can everything else be subordinated.

With that as The Mission, it is very hard to argue that any other activity should take precedence over it. Even in peacetime. And the brass at NWC, no fools they, play upon this sense of military obligation and duty—remember, this is a military base and almost all the staff, students, and personnel are military—in order both to get things done and to shape-up the civilian faculty. For in the name of accomplishing The Mission, you can justify just about anything.

Here are some instances of how the idea of The Mission can be both used and abused:

- I was told by the brass on several occasions that, since my research was not directly relevant to The Mission of the University, I should desist from such activity.
- Mission Success was also used to rationalize obligatory attendance at ceremonial occasions, sports events, extracurricular lectures, etc.
- Mission Success was similarly employed to justify extra teaching assignments, guest lectures—almost anything that the commandant wanted us to do.
- "Completing the mission" was used to oblige the faculty to serve as hosts, mentors, housing specialists, and family counselors when international students arrived in Washington.
- The success of The Mission was used systematically to stifle faculty dissent, close off discussion, and keep the entire staff focused on a single goal.
- Who can argue with The Mission and certainly not in public? That *was*—the training of military officers—our prime mission. One might suggest that there could be other missions, too, or that there might be alternative ways of viewing the primary mission. But when the Brass—department heads, deans, the commandant—invoked The Mission, the rest of us were usually subdued into both silence and acquiescence.

STUDENT-ORIENTED

I've never seen an institution that is as student-oriented as the War College. *Everything* is done of, by, and for the students. It's the exact opposite of every other academic institution with which I've been associated where the faculty, generally, rules. Not so at NWC; there, it is the students who rule. Or, as we like to say, where the "inmates" are running the "penitentiary."

The origins of this system are difficult to discern. Certainly that was not the plan of the College's founders, including George Kennan who conceived of a school not very different from a civilian university where the faculty did original research and writing and taught, and the students learned from the faculty. But over the course of the decades since Kennan, the faculty's research role has been diminished almost to the vanishing point, while the position of the students has been augmented. Why this reversal of roles?

Part of the answer lies in the fact that the military brass at NDU and NWC do not understand and see no reason for its civilian faculty to engage in original research work—a topic discussed in more detail in the next chapter. Instead, and the reverse side of that coin, they see NWC as *exclusively* a *teaching institution,* which, of course, makes the students the focus and not the faculty or their research efforts. In addition, the students have learned over the years to manipulate this focus to their advantage. The students we teach (and the students know this) will be the leaders and flag officers of the next generation; they will soon outrank, at least in the military hierarchy, their former teachers at NWC and be in a position to do something about it. So if you're an administrator or faculty member at NWC, you'd better treat the students nicely or else two things will happen: (1) the students will go to their friends and colleagues on the staff of the Joint Chiefs, under whom NWC operates, and bad things will then happen to the NWC budget or present personnel, and (2) the students will wait until they become generals and admirals (generally, only a year or two after completing NWC) and mete out their revenge then. Witness the case of our own commandant, a former student at NWC, who waited fifteen years to wreak his revenge on NWC for some real or perceived faults.

Here are some instances of how this works at NWC, the cumulative effect of which is to add up to a strong case:

- When I first arrived at NWC, one of my colleagues took me aside and told me the College is "overly student-oriented." He said, "When the students complain, things happen right away; when the faculty needs something, it may take years."

- My former dean, who should know better, was forever waxing eloquent about "how good the students are" and "you'll love the students." These *are* high-flyers but, having seen them up close, taught them, and graded their papers, I wasn't that impressed; they're not as good as my own civilian graduate students.
- The commandant was seeking to fill the prestigious Kennan Chair with a distinguished foreign policy specialist. When he learned I worked with Zbigniew Brzezinski at CSIS, he called me into his office to see if I could help with recruiting him. In the course of our conversation, the commandant remarked, "Well, he would come because of the student body; the faculty does not have the reputation to attract him here."
- Course 4, on Geostrategic Context, used to divide at its midpoint after four weeks, and we faculty would get a whole new roster of students in our discussion sections whom we didn't know and had never seen before. The shift made no sense pedagogically; but militarily and from the students' viewpoint it did. This way, I was told after inquiring, the students got to know *each other* better and camaraderie was built.
- Much the same holds true for discussion sections in general. In civilian colleges, discussion sections are usually held only in beginning courses; but at NWC, at the graduate level, discussion sections take place every day. I was told, since not much enlightenment comes out of these discussions, that they were solely to give the student practice in, and the opportunity of, speaking up in public. In short, enlightenment was not the issue, speak-practicing was.
- One of the students was caught turning in the same paper for two different classes. I got wind of this because the paper had originally been submitted in my class and received a good grade; on that basis, the student decided to submit it again! In civilian universities, this is a no-no, a grounds for student judiciary procedures or perhaps even expulsion. But not at NWC. There you'd have to almost literally murder someone in order to be expelled! The student got off easy: his own lawyer from the judge advocate's office, no expulsion, no judicial procedure, nothing in his permanent file (which in the military can be a career ender), only a little verbal slap on the wrist—tut, tut—and the requirement that he produce a new paper.
- One day when I was not in my office but the door was open, a student came in and "borrowed" one of my own personal books. When I found out about it, I was incensed. In civilian universities, first, you don't enter an office without being invited; second, you *never, ever* borrow something without asking. But when I complained to the dean, the commandant got involved and it now became my fault. In response to the issues raised above: why wasn't I in my office when the student needed me; faculty offices don't

belong to the faculty and are not sacrosanct, they are the common property of the College and open to all; and books in faculty offices are for *students* to use. Wow! How do you like that for an interpretation? At least I eventually got my book back.

- Students at the College were pampered and every effort was made to accommodate *their* needs. Have I remarked that the class president had his own reserved seat in the auditorium and—better than mine—his own reserved parking slot?
- Over the period of my time at NWC, the trend was for the students to choose not only among a broad array of electives but also to choose the *time* and *place* of the course. The faculty was expected to accommodate these demands by teaching more sections at the time the *students* chose.
- As my department head once inelegantly put it at a faculty meeting: "There's only one reason for your (the faculty's) existence here: the students!!" There it was very baldly put; the students are the *only thing* that counts.
- One of my students had been absent from class for several days, but I'd seen him in the building. And so I'd inquired why he wasn't in class. The answer was that he was on the College bowling team. And bowling helps build class spirit and unity, valued at NWC more than anything learned in the classroom. So the student was excused. But not my colleague Hugh De Santis who was told he could not miss a half-day to brief the Joint Chiefs on a critical issue, NATO, Hugh's specialty.
- I had a very moving final meeting of my Course/seminar. The students were very grateful, gave me a round of applause, and came up to shake my hand afterwards. They'd learned something and were appreciative. They even presented me with a spectacular gift: a piece of barbed wire from a part of the Iron Curtain, enclosed in a bloc of Plexiglas, engraved, and wrapped in a red ribbon. Since the Cold War had been so much a part of my life for forty-five years, and the gift was given with such appreciation and sincerity, it was a moving experience. Maybe teaching at the War College, despite the difficulties and perverse priorities, is not so bad after all.

MILITARY COMPLEXES

Once I'd been there for a while and come to understand the institution better, I began to realize that all that military bluster, the jealously guarded command system, the almost hyper emphasis on rank, and the monopoly maintained over all the key administrative positions at NWC were all just expressions of military inferiority complexes. It was not that the command system and the structure of top-down, hierarchical authority reflected good order and the strength of the institution; rather, that was used to disguise weaknesses, insecurity, incompetence, and the inability of most military officers to function in a non-military environment. The reason the NDU president, the NWC commandant, the deans, department heads, et al. kept such tight control is, not that they were powerful, but instead that they were *afraid,* afraid the civilian faculty would show them up, come to dominate, and reveal to all how inadequate the military running of the War College really was.

One of my first commandants was chock full of such complexes. He would turn visably nervous when in the presence of the civilian academics at NWC. He would seldom deal with the academics directly but usually hid himself in his office, barricaded behind his secretary's forbidding presence, and usually spoke to the academics only through his XO and the dean of faculty. When he did address the faculty, he was blustery, smart-alecky, and defensive. Thankfully, he was also mercifully brief, eager to get back to the defensive perimeter of his office or to his tennis game.

The commandant was perhaps an extreme case, but many of these same traits were often present among the deans, department heads, and the military faculty. Their bluster, the command system they enforced, the almost constant efforts to embarrass the civilian faculty and put them in their place derived not from strength and self-confidence but from weakness and a lack of confidence. The military faculty had no records of research; they lacked experience in the classroom; and they lacked knowledge on the literature, theory, and subject areas they had to teach. Often they sought to bluff their way through a discussion section and thus received bad student evaluations. Or they covered over their own lack of knowledge by letting the students do all the talking, thus abdicating their role as teacher.

It was this commandant after all, seeking to compensate, who proclaimed that "every officer is a teacher." Well, yes, of course, in a very limited way, but that could not cover over the lack of expertise or the part of the commandant himself and most of the military faculty. Such expertise is required to function effectively at what NWC purported to be: a graduate-level program. When the bluster didn't work, various plans were put forth to have the civilian faculty mentor the inexperienced military faculty, to meet with them, to have them sit in on our classes—as if that brief exposure could bring inexperienced teachers up to acceptable levels. As if you could convey years and even decades of knowledge and experience in a single, brief session. When this kind of "instant" teacher-training programs didn't work at bringing the military faculty up to acceptable levels, the opposite tactic was frequently applied: to denigrate and bring down the civilian faculty to the mediocre level of the institution as a whole. Both kinds of behavior are on display in the following anecdotes:

- The armed forces have an absolute and understandable prohibition against outside income; at NWC, that had been extended to the civilian faculty. But that cannot work: we have income from book royalties, manuscript reviews, faculty evaluations, and other sources; in addition, we are frequently asked to give outside lectures for which honoraria are provided. So when one of these opportunities came up in my first weeks at NWC, I went to see XO Col. Kirk Lewis. He was very accommodating about it, recognizing that the civilian faculty needs these kinds of outside professional activities, and ruling in my favor. In the course of this meeting, Lewis made it clear that he wanted to retain NWC's civilian faculty (occasionally there had been rumblings from the commandant of moving to an all-military faculty—simpler that way), and that he recognized full well NWC could not attract and retain high-profile academics unless this ban were relaxed. I came away from this meeting convinced the Army wants desperately to retain control here, is deathly afraid of ceding any control to the civilians, but to compensate they will do almost anything to keep the civilian faculty here and contented.
- In the span of one week, (1) the commandant's secretary goes on vacation so he commandeers our (my!) departmental secretary, without letting us know beforehand and with no sense that she may have other work to do; (2) we get a memo indicating—no consultation, no by-your-leave—all trip directors *will meet* (that's an order!) on such-and-such a date; (3) the commandant, without request or permission, sits in on my class, thereby ending, with him there, all possibility for real student-faculty discussion. From the military's point of view, that's his role as commandant, to check on us, to make sure we're not like the proverbial "drunken sergeant at the gate." For the military, this is Standard Operating Procedure (SOP); for civilian faculty, this kind of "spying" in the classroom is unacceptable. I chalk it up to military (or the commandant's) ignorance of how classroom learning occurs but, more than that, it also reflects insecurity; hence, the command's need to constantly assert authority, and also a fundamental distrust of the civilian faculty.
- The Defense Department will not pay for senior officers to get a Ph.D. since they're already at the end of their careers; yet junior officers who acquire advanced degrees (and the expertise that goes with them) are not listened to by their seniors simply because their rank is too low. At the same time, full colonels on the NWC faculty who have Ph.D.s are passed over for promotion to flag officers because they lack experience in deployable bases. It is

an absurd and self-defeating system that robs the armed forces of expertise. My understanding is that, since Iraq and Afghanistan, the Pentagon is finally, after sixty years, in the form of a revised Foreign Area Officer (FAO) program, coming around to recognizing the value of regional, language, and country expertise.

- The then NDU president has just promulgated a document on civilian faculty. Of course, this was done in his capacity as commanding officer; no consultation with or input from civilian faculty was involved. The document centralizes *all* power in his hands; it overrides all previous documents that attempted to spell out faculty rights; and it gives the president complete authority to fill the "needs" and carry out "the mission" of NDU as he defines it. There are in this document no protections, no guarantees, no academic freedoms, and no redress of grievances. The NDU president's "model" is that of a military base; there is no bow whatsoever to the fact that NDU is simultaneously a civilian teaching institution. Our (civilian faculty) interpretation is that the president is preoccupied with control; he has authority hang-ups. He also informs us he is terminating all discussion of tenure because tenure also means a loss of his control. He wants to run a tight ship and is fearful of anything slipping out from under his control. Like our own commandant, he's intimidated by, fearful of, and wants no contact with the civilian faculty. Clearly, he's the wrong person for the NDU presidency, which requires *both* a military and a civilian-academic model. Fortunately, his decrees are never implemented.
- As part of this same control regime, the NDU president's office will now decide whether or if we faculty members can be interviewed, quoted, or appear on television. Even on issues in our areas of expertise. This is part of the Public Affairs Office at the Pentagon's efforts to make sure all expressed employee views are in accord with DOD policy. NDU's president, of course, sides with the Pentagon and against the wishes of the faculty. Fortunately, this provision is not consistently implemented either, but it nevertheless hangs like a club over the faculty's heads, discouraging public and media appearances. Which, of course, in the academic world are viewed as absolutely necessary as a way of testing your ideas against others'.
- The Air Force is insisting that our then dean retire. The new commandant has in mind as his replacement a weak, compliant, milquetoast person who will be completely subservient to his control. I do not think this is a good idea and go to see former dean, Air Force Colonel Roy Stafford. Roy informs me there's not much we can do; this is a "command decision," and the command will not countenance *any* faculty say in College governance. Again, I ask, "Is this because of the command system or is it for fear of the civilian faculty and the loosening of military authority?"

- One of our faculty members was put on "probation" and offered only a two-year contract instead of the usual three. His transgression was to ask a student to preside over his class while he did a National Public Radio (NPR) interview. Not exactly a major transgression. This faculty member is also viewed as "uncooperative" because he doesn't attend all NWC training sessions, which are mainly for the military faculty. It's a big to-do at NWC. My interpretation: it's another control/complex issue. NDU is afraid of loosening the screws; it's afraid of losing control; it needs to keep its faculty under lock and key.
- A report filters down from the commandant via the deans and department heads that, even though the size of the student body is staying the same and that of the faculty going up, our teaching loads must also go up. That makes no sense; how can it be? It is because none of the new military faculty coming in has any teaching experience, and we must fill in for them. Apparently, NWC, which once could select among the military officers assigned to it (i.e., it looked for teaching experience or at least some measure of academic background), had lost that ability. The result, on the old and discredited assumption that every officer was a "teacher," was too many military faculty assigned to the College who not only had no teaching experience but no intellectual curiosity either.
- Dean Roy Stafford reports to us that his proposal for faculty merit awards has died. It had no support from the military hierarchy and was vetoed by the commandant and the NDU president. The command argued (1) it had no way to measure merit among the civilian faculty; (2) merit pay increases would create discrepancies and, hence, anger and envy among the faculty; and (3) since the uniformed services had no provision for merit *within* a given rank (it is an up-or-out system), civilians should be governed by the same rules. Of course, the non-productive faculty supported this ruling since they would be rewarded regardless of performance; I thought it was a military cop-out.
- Department meetings at NWC are rare. My notes record how surprised I was when one was called. Department meetings at NWC do not, as at a civilian university, involve lively give-and-take; rather, they consist of announcements and orders from the commandant or the NDU central administration, via the department head, to the faculty. In keeping with the command system, it is purely a one-way communication. There is no faculty input; I don't recall a faculty vote ever taken on anything at one of these meetings. Rather, it is a means by which orders are conveyed to us without calling them "orders."
- Our current department head is increasingly intensive. He's worse than a military commander, "little Napoleon." He's become like the "soup Nazi"

on the Seinfeld television show. He's constantly looking over our shoulders in our offices, wants to sit in on our classes, sends spies into some classes, and writes up his own evaluations of our performance without sharing those with the faculty. He'd previously vetoed my research agenda, prohibited the secretaries from typing my scholarly papers and articles, and ruled out the mailing of professional letters and materials as "private" activities. I'll be very happy when we can get rid of this guy.

- The faculty has been marginalized. Toward the end of my tenure at NWC I'm terribly discouraged. I see no way to plug into the system or to change it at any level. At every position within the hierarchy, one meets brick walls and/or incompetence. The NDU president proves to be a huge disappointment: authoritarian, arbitrary, unapproachable, centralizing all power in his own hands. The new commandant is not dumb but he's inexperienced in education, a waffler, indecisive, not helpful. The dean is another disappointment: lacks confidence, doesn't meet with the faculty, in over his head, not up to the job. Incompetence and the wrong people at all levels.
- We get a threatening memo from our department chair telling us we must play host to the international fellows. We're obliged to set aside two-and-a-half weeks when they first arrive in August, helping them and their families settle in, adjust, find housing, and get acclimated to American culture. Sorry: I'm willing to work with the international fellows *in the classroom;* I'm not willing to play nursemaid, house-hunter, and baby-sitter. It's completely unprofessional. If you want to help the international fellows, fine, then *hire* a housing specialist. But in the military culture, this is interpreted as lack of dedication to "the mission," if not insubordination. The faculty's reaction to this memo is so negative that the department head backs away. Yeah, a victory (one of the few) for us!
- The new department chair apparently keeps a gigantic matrix by which he's evaluating all the civilian faculty. Not just on our teaching but on our general comportment. Hence, if we oppose his visiting our classrooms or fail to assist the international fellows, we're marked down as "uncooperative." Any little transgression or sign of unwillingness to assist gets us a black mark. Most of us are convinced the chair's enthusiasm for policing the faculty is his own but we're suspicious that he's also acting on orders from on-high to "shape up the faculty." It's a disaster; the atmosphere at NWC is shifting from "merely" authoritarian to totalitarian. We cannot function this way.
- In April 1996, I went off to Portugal and Spain on a research trip. It was *so* nice to get away from the suffocating environment at NWC. But immediately upon returning, I run into the new associate dean, a civilian (that's a first), Ron Tammen. I mention to him that, if we're going to have this

elaborate faculty review that's now being announced, we must have (1) clear criteria, (2) the purposes of the review clearly spelled out, and (3) confidence in the process. None of this presently applies. To my surprise, since he's now part of "the command," Ron agrees. He also confirms my suspicions that this process is intended to rank-order the faculty. Two of the best scholars among the faculty are on the bottom of the totem pole. The commandant wants to use these rankings to create "openings" that he can fill. The offending faculty members are then brought in for an oral tongue-lashing; eventually, they will be let go.

- My own time for evaluation, along with that for the rest of the faculty, comes. I have to meet with the department head. He says I spend "too much time" in research and writing. I remind him of George Kennan's injunction that NWC must be a place both for research and teaching; it has no effect. We then turn to my teaching: 85 percent of my students rate my courses as "excellent"; the rest, as "good." You can't get much higher ratings than that in a classroom, but the head wants to gloss over that. He focuses on my travel and asks why, in Japan, my institutional affiliation for the upcoming International Studies Association meeting is listed as University of Massachusetts and not NWC. I'm surprised that he spotted this, but he's an ISA member. And actually, even though I'd not told that to the head yet, in the spring of 1996 I'd just been awarded the Leonard Horwitz endowed chair, a hefty pay increase, a research assistant, and yearly guaranteed travel funds, and was already planning to go back to my home university, UMass. In the next couple of weeks I would let the head and everyone else know about my plans. And that it was not just the pull of the UMass offer but also the push of the stifling atmosphere at NWC that drove me to accept.

CONCLUSION

There are many other aspects of "the military mind," as it was manifested at the National War College, upon which we could comment. Gender discrimination remains a problem: in my department, National Security Policy, there was at first one, then two women faculty out of twenty-two, or under 10 percent; in the Department of Military Strategy and Operations, there was one woman, the colorful, Israeli-born, strong-willed Ilana Kass, and eventually two. Racial discrimination is similarly a problem: in neither of the two departments were there any blacks or Hispanics, and at senior officer (lieutenant colonel and above) ranks these two minorities were exceedingly scarce, both among administrators and the student body.

The military is very team-oriented. At several places in this chapter we have seen that building camaraderie and team spirit through bowling, softball, and other activities was deemed more important than attending class. "Unit morale" is another important military virtue—again important for taking San Juan Hill but not necessarily for the faculty at the National War College where dissent and petty squabbles, often aided and abetted by the actions of the brass, seemed endemic. "Tasking" similarly meant carrying out orders without any input from the person tasked.

The War College makes a big deal about "going by the book." That's common in terms of armed forces' rules and regulations. The trouble is, there is no real "book," in the sense of rules and procedures, to govern the civilian faculty at NWC. And my experience was that it was the civilian faculty, hoping as we see in more detail in the next chapter to get clarification on such things as research, tenure, and academic freedom, that most often was clamoring for guidelines and procedures, which the military brass failed to provide. The only conclusion I could reach from this is that, when the deans, commandant, or NDU president didn't want you to do something, they would haul out "the book," in the form of obscure military regulations, that seemed to ban you from doing it. On the other hand, when it was something the civilian faculty wanted to get regularized, that was consistently viewed as interfering with the commandant's "command" responsibilities. You could not "win" or get much accomplished this way. Students of bureaucratic behavior will find all this quite familiar.

To answer now my initial question: Is there a military mind? The answer is, resoundingly, "Yes," even at the National War College. In earlier chapters, we had seen that the armed forces officer corps is, overwhelmingly, disproportionately, Southern, small-town, conservative, Republican, and Christian. Those features powerfully shape the military's way of thinking, its political culture. Then if we add in those features that are unique to the professional military institution—the emphasis on order, authority, hierarchy, rank, discipline, command, mission, etc.—we have even more characteristics of the military mind. Taken together, both the sociological, religious, and political features referred to initially, and those that accrue through professional and forces training—these features add up to a quite distinct and unique military political culture.

The question now becomes: Is this military political culture or "the military mind" compatible with a free-wheeling, liberal, academic, political culture that questions all revealed truths and the authorities who seek to promulgate them. In the next chapter, shifting gears from this chapter's focus on military values and how they apply at NWC, we focus on those issues that are of most concern to academics: academic freedom, tenure, research and publications, teaching loads, sabbaticals, the right of dissent, the academic climate. After this survey we will be in a good position to discuss, in the Conclusion, whether the military and the academic cultures can ever be bridged, made compatible, and can live in harmony. Or not.

NOTES

1. I have visited, lectured at, and explored this issue at all three military academies (Army, Navy, Air Force), the Army War College, the Navy War College, the Air War College, the Navy Post-Graduate School at Monterey, California, the Command and General Staff College at Fort Leavenworth, Kansas, as well as at NDU/NWC.
2. See Richard Betts, *Soldiers, Statesmen, and Cold War Crises* (Cambridge, MA: Harvard University Press, 1977).
3. C. P. Snow, *The Two Cultures* (Cambridge: Cambridge University Press, 1998).
4. Huntington, *The Soldier and the State: The Theory and Politics of Civil-Military Relations* (Cambridge, MA: Belknap Press of Harvard University Press, 1957).
5. In my academic career, I've written extensively on authoritarian regimes. So I know how these systems work. See Howard J. Wiarda, *Dictatorship and Development: The Methods of Control in Trujillo's Dominican Republic* (Gainesville: University of Florida Press, 1968) and Howard J. Wiarda, *Corporatism and Development: The Portuguese Experience* (Amherst: University of Massachusetts Press, 1977).

Chapter 7

The Academic Mind

The academic profession and the academic mind are very different from the military profession and the military mind. Whereas the military profession emphasizes order, discipline, and the unquestioned carrying out of orders, the academic profession is skeptical, irreverent, questioning, and unwilling to carry out orders unless their logic and good sense are explained. Whereas the military profession emphasizes hierarchy, authority, rank, and command, the academic profession sees no merit in those features whatever, is skeptical of all forms of authority (deans, department heads, provosts), and accepts scientific breakthroughs, if valid, even from the lowliest graduate students. The reasons for these fundamental differences, of course, are that the purposes, assumptions, and goals of these two professions are very different and, as might be expected, are often at odds.

But the differences go much deeper than these professional and career orientations. They involve religion, culture, society, and politics as well. The military ranks tend to be religious while the orientation of scholars is scientific and secular. Military men tend to be cultural conservatives in their preferences on art, movies, or music while academics tend to prefer progressive and *avant garde* materials. The troops oppose gays and women in the military, while academics are overwhelmingly favorable on these social policies. And while military officers are 80 to 90 percent Republican, academics are 80 to 90 percent (or more) Democratic. On a *host* of issues, the professional military is much more conservative than the rest of the population, whereas on these same issues academics are much more liberal.

The question, as posed at the beginning of the previous chapter, is whether these two "worlds," these two "cultures," can ever be bridged and reconciled? Can the Department of Defense Professional Military Education (PME) schools ever be comfortable with all these liberal academics within their midst? Conversely, can civilian academics ever find happiness (it sounds like a soap opera and, in part, it is) in working at a military institution? These questions are at the heart of the discussion in this chapter; in the Conclusion, we return to these themes and seek to push the analysis even deeper.

ACADEMIC FREEDOM

I think of myself as a centrist politically, maybe center-right now that I'm in my emeritus stage. But I also think of myself as a serious scholar and academic, and certainly not an ideologue. One might think—and I certainly did initially—that mine would be a good appointment at NWC: a serious academic whose social and political views are not so far distant from those of the military brass and of the lieutenant colonels and colonels whom we teach.

But as a serious scholar I'm also committed to academic freedom. By academic freedom we mean the ability to explore all subjects, to speak our minds about them, to write and publish on these subjects. In the classroom that means we have the right of free speech, the ability to speak truth to power, regardless of the consequences. Particularly within our range of expertise, but on other subjects as well, our voices cannot and should not be silenced. We can research any topic, speak out on the issues, and publish scholarly works as well as opinion pieces on the main topics of the day. That applies particularly to foreign affairs, international relations, and foreign policy where *all* the NWC faculty have specialized knowledge and expertise.

But this university-based conception of academic freedom is not how it works at the National Defense University:

- First, I discovered that NDU and NWC had almost no rules, procedures, or processes in place to protect academic freedom institutionally. There was no system of faculty governance, no grievance procedure, no process by which faculty could make their voices heard in decision-making. That was, it turned out, a conscious policy: it helped keep the civilian faculty weak and powerless; it also preserved the hallowed military principle (remember, NWC is an Army base) that the commandant was in *command,* could order people around, and initiate and terminate people or programs at will.
- Second, I discovered that, while not officially proclaimed as such, NWC had an official position on almost all issues. Those positions were largely contained in the lengthy, elaborate, course syllabi from which we taught. The readings were prescribed; the arguments and main points were listed; and we proceeded by deductive reasoning from the text. There was no room

for originality, to build in the results of our own research (I did so anyway), or to deviate from the official position. And, of course, nothing we said or wrote could be allowed to deviate from the official DOD position. That is not my definition of academic freedom.[1]

- In 1993 a new Academic Review Board was instituted at NWC. It was designed, so we were told, to provide an avenue for faculty input at the College. But over half the members of the Board were military and from the administrative side, and it had no independent standing. So far as I know, the Board was never convened and never functioned as its statutes stated.
- The NDU president one day announced that he was promulgating a document dealing with the civilian faculty. The process was entirely top-down; there had been no participation by the civilian faculty whatsoever. The document that the president promulgated concentrated just about all power in his own hands. Decisions once made at the NWC level and thus close to the "troops" (us) were now taken away and lodged at the NDU level. The basic message of this document was that the NDU president could ignore all previous commitments and policies at NWC and himself make all decisions on personnel and other matters based on the "needs of NDU," as he would define them. There were, in this document, no protections, no guarantees, and no means for redress of grievances.
- When I'd had my initial job interview at NWC and then after arriving there, I was repeatedly assured that academic freedom was "absolutely" protected. But that did not prevent my being "tasked" to lecture to the RCNSC, the Pinnacle class, or the wives' group. I quickly realized we would be "tasked" for other extracurricular activities as well, *required* to attend ceremonial events, and deemed to be "on duty" at all times. None of these correspond to my idea of academic freedom.
- NWC can be an intimidating place. It was especially so under one of my first commandants. This commandant was "old school," jealous of his prerogatives, and both defensive and complexy about them. All of us learned quickly that we should not cross or even question him, or bad things could happen. Again, that atmosphere, not entirely unheard of with some civilian university deans and department heads, is hardly in keeping with academic freedom.
- Colonel Dick Haney, an Army officer but housed in my National Security Affairs Department, was asked to go on the prestigious nighttime television program "Nightline" to speak on his specialty, peacekeeping in Bosnia. But the Public Affairs Office at the Pentagon refused to allow him to go on. And, as a serving officer, Haney had to obey. Some of the civilian faculty, however, were less reticent about what we spoke or wrote, or on what programs we appeared. Nevertheless, whenever we did so, the sword of Damocles always hung above us; we were constantly worried that Public

Affairs would clamp down on us, too. It was not a climate conducive to academic freedom.

- When he promulgated his program of centralized authority and decision-making, the NDU president called the faculty together (after the fact) to indicate that he also believed in academic freedom. But *"of course,"* he said, that could not include "anything" that was contrary to Department of Defense policy. And that is the way I came to understand academic freedom at NWC: it was a principle honored in the abstract; but if it came to a choice between academic freedom and NWC or DOD policy, or if it clashed with a commandant's authority, it was academic freedom that would have to be sacrificed.

SABBATICALS

NWC has, *in principle,* a system of sabbatical leaves, but sabbaticals are not viewed there in the same light as in universities. In universities, especially research universities, sabbatical leaves are viewed as essential to the research function of the institution. They are seen as entitlements, more or less automatic after you've filled your required years of obligation to your university. From time to time, faculty members need to get away from teaching and administrative work to devote full time to their research. Such research is valued highly by the institution, adding to or perhaps *the* key component in its prestige. In my own university career, the rule was that you were entitled to a sabbatical once every seven years. That is, you taught first for six years and, at the end of that period, you were *entitled* to a sabbatical. In that seventh year sabbatical, you could either take the full year off from teaching at half-pay or half of the year off at full-pay. The assumption was that, during your sabbatical, free of teaching and administrative responsibilities, you could produce several scholarly articles and, hopefully, a book. Nowadays, universities are more flexible about sabbaticals, often offering junior faculty time off with pay for research purposes after only three years of teaching and service.

NWC has a section on sabbatical leaves in its policy manual for faculty members, issued as NDU Regulation 690–4 (1 May 1991). Sabbaticals are formally "on the books." The language is quite expansive, indicating that sabbaticals may be authorized with full pay for up to one year—a better arrangement than at most universities. The NDU president is authorized to grant such sabbaticals, which are specifically for advancing the educational mission of the University, not for research. A faculty member granted a sabbatical may not take on other full-time teaching obligations during the period of the leave and owes the College two years of service in return for the sabbatical.

But in reality, sabbaticals at NWC are only rarely granted. They are not a normal, regular part of the process. The regulations are very clear that the sabbatical is not an entitlement. And there are numerous grounds upon which it may be denied. For example, when I first got to NWC and inquired about sabbaticals, I was told that the commandant was opposed because he thought it left the military faculty, who are not eligible for sabbaticals, feeling like second-class citizens. Next, in a faculty discussion of sabbaticals, the commandant told us that he needed to balance off the civilian faculty's desire for research sabbaticals with the military faculty's desire for leaves to go back to graduate school to earn an advanced degree. If such a program for the military faculty were to be put in place, he said, then the civilian faculty might have to wait nine or ten years for a sabbatical, instead of the usual six, so that the teaching obligations of the College would not suffer.

The National War College does not see the value of sabbatical leaves in the same way as do research universities. First, it thinks of itself as primarily, even exclusively, a teaching institution. It has largely abandoned and sees no value in the original George Kennan conception for NWC, as a teaching, research, and contemplative institution. Second, like many military schools, NWC teaches "from the book." That is, it teaches from established texts, from "revealed truth," and its method is deductive. Therefore, it sees no need for any original research on the part of the faculty. And if there is no need for research, then why have sabbaticals, whose whole purpose is to free up the faculty for research? Except perhaps for a very limited sabbatical leave program, which is what NWC has, solely to keep the civilian faculty contented.

Once again the best way to make this point is via a series of anecdotes:

- My first commandant at NWC, undoubtedly influenced by an earlier, pre-Title 10 conception of the College, was convinced the faculty was not very good. He often belittled the civilian faculty and once told me he thought the students were better than the faculty. If the faculty was of poor quality, he continued, why should we have a sabbatical leave program to support what would certainly be poor and not useful research?
- In my first year at NWC, East Asia expert, Paul Godwin, led the charge to institutionalize sabbaticals. He sought to make the case that sabbaticals would be used not only for research but also to develop new courses. That sabbaticals, presenting the case in a way that would appeal to the commandant, would improve teaching at NWC. But the commandant was not buying the argument. He saw no value in faculty research or research leaves. Only reluctantly did he agree to continue the existing "informal" system of sabbaticals, but he wanted no part of further institutionalizing the system.

- The question then came up about sabbaticals for the military faculty. Could NWC put in place a program that enabled senior officers to go back to school to get a Ph.D.? But the commandant responded: (1) that would require a three-year leave (beyond the MA), which NWC could not afford, (2) it would interrupt an officer's career path and result in his being passed over for promotions (the up-and-out system), and (3) to satisfy the teaching needs at NWC, if we allowed multi-year sabbaticals for the military faculty, it would mean the civilian faculty could have a sabbatical only every nine or ten years. I think this last response pretty much ended any talk of military sabbaticals at NWC.
- But then things got worse. The commandant told us he was very uncomfortable with the idea of sabbatical leaves for the civilian faculty because that left the military faculty, who were ineligible for such leaves, feeling like second-class citizens. At this stage, the commandant had us coming and going: no military faculty leaves for the reasons given above, but no civilian faculty leaves either because it made the military guys feel bad!
- By 1995 faculty member Paul Godwin, who'd been at NWC the requisite six years and was, therefore, theoretically eligible for a sabbatical, applied for a leave. But his request was turned down. The brass at NWC told him he was needed in a teaching capacity and, therefore, his sabbatical would have to be postponed indefinitely. The decision was wholly arbitrary; there was no recourse. The lesson this decision brought home to us was that so-called "earned" sabbaticals are not at all automatic; they can be postponed or denied on almost any grounds, in this case that of "teaching priorities." The decision in this case, top-down as usual without any faculty input, reflected once again the non-institutionalized nature of procedures at NWC and the corresponding ability of the commandant to make "command" decisions unhindered by institutional procedures.
- The Godwin case had a profound effect on me and my future. For I was then in my fifth year at NWC, increasingly unhappy in my position there but willing to stick it out if I would receive a sabbatical leave after my sixth year. But if sabbaticals could be turned down so arbitrarily, why stick it out at NWC in an unhappy situation if in the end my "earned" sabbatical could be capriciously denied. The denial of a sabbatical to Godwin hastened my own decision to leave the College sooner rather than later.[2]

RESEARCH

The National War College places almost no emphasis on research. It honors research and publication in the breach, in theory, but it sees no relevance for the research that the faculty does to its primary mission, which is teaching.

NWC is preeminently a *teaching* institution. In this respect, it is not unlike many colleges in the United States that pride themselves as primarily teaching institutions. The difference is that these are undergraduate teaching colleges, whereas NWC is a graduate institution. In fact, it is exclusively a graduate institution. *All* of its students have at least a bachelor's degree. How can you not emphasize research in an institution that has only graduate students? Something has to give here: either NWC has to emphasize research as well as teaching in its capacity as a high-level (the highest of the PME schools) post-graduate program, or it has to admit that it is really teaching at an undergraduate level. Either the one or the other; you cannot have it both ways.

I am convinced that, on research, the military, or DOD, just doesn't get it! In this respect, they are not unlike other laymen in the educational field. They fail to see that, especially at the graduate level, research and teaching are and must be intimately related. Good teaching must be informed by good research. The best teachers are usually also the best researchers. They are good teachers precisely because they are in the forefront of research. Their original research imparts not only innovation and the latest research findings to their teaching but also infuses their teaching with enthusiasm and excitement. And that excitement about learning is then transferred to their students. True, there are occasionally exceptional teachers who have wonderful communication skills who are not actively involved in research. But these are few and far between. Generally, the rule holds that the best teachers are precisely the ones who also do exciting and innovative research. That is precisely why we send our sons and daughters to the nation's great research universities such as Harvard, Stanford, Berkeley, Michigan, and many others.

But the military mind does not understand this connection between good research and good teaching. It is like a lot of parents who similarly want to send their children to good "teaching colleges" but who fail to comprehend this research-teaching link. Hence, at NWC there is almost no emphasis on research. In fact, the College actively discourages research by its faculty. It wants us to teach from "the manual"—i.e., from a prescribed syllabus. You have a manual for tank-driving, so why not something similar for foreign policy and national security. That is why all the core courses and electives at NWC have such large syllabi, upwards of three hundred pages. Everything is laid out: the readings, the concepts, the conclusions. Nothing is left to "chance," let alone to my original ideas on the part of the faculty.

The method here is deductive rather than inductive. Everything that is worth knowing about the subject is laid out in the texts and the syllabi. You learn by rote from "the book," not from the excitement of new and innovative research findings. Non-prescribed readings and those that depart from the accepted wisdom are not permitted. In other words, we teach using the scholastic method: there is a prescribed "truth"; we teach that, and there is no deviation from it—let alone any room for incorporation of our own original, derived-from-our-own-research insights.[3]

Not only does the military brass not "get" this indispensible link between good research and good teaching, but there is more to it than that. The brass has, I concluded while at NWC, almost no respect for its own civilian faculty. Part of this attitude stems from the legacy of the earlier, sorrier, pre-Title 10 low quality of the faculty in the 1970s. Part of it stems from the military suspicion that we civilian professors are all lefties, long-hairs, Vietnam War protestors, and bra-burners. And part of it also comes from a general military disdain for civilians, who are viewed as "soft" compared to the professional military and who, in their view, have not gone through the same rigors career-wise as have the officers.

The reality is that the new faculty at NWC are as good as at any program or institution of international affairs anywhere in the country, that most of us on the faculty are quite moderate politically and not wild-eyed war protestors, and that I will stack the rigors of my career tests—GRE exam, graduate courses, field work in the Third World, MA thesis, more seminars, Ph.D. comprehensive exams, another year of field work, doctoral thesis, job search, assistant professorship, tenure, promotion, etc.—against those of military officers any time. Regardless of these realities, the military hierarchy at NDU/NWC continues to be disdainful of the civilians on its faculty, to treat them as second-class citizens, and not at all to value their research and writing.

All this is a very long way from George Kennan's original 1947 plan and conception for the College. Recall that Kennan and his early colleagues were committed to creating an institution dedicated to research, advanced strategic thinking, *and* teaching. It was at NWC that Kennan had written the famous "Mr. X" article that outlined the entire U.S. Cold War strategy; wrote the book, *Measures Short of War* (based on his lectures at NWC) that explored in pioneering fashion the relations of war, diplomacy, and other "soft power" strategies; and formulated many of the ideas that went into the organization of NATO, the Trans-Atlantic Alliance, Point Four, and other innovative policy decisions. But after Kennan and his team left NWC, the College floundered for several decades and virtually the entire strategic research focus was lost. NWC and its students were the great losers in this transformation.

When I arrived at NWC in the early '90s, I found a mixed situation as far as research was concerned. About five or six of the new hires, all Title 10s—Mel Goodman, Hugh DeSantis, Richard Melanson, Dale Herspring, Marvin Ott, myself—had an active research agenda. Several others, who had been hired earlier—Terry Deibel, Bard O'Neill, Paul Godwin, Ron Tammen—produced maybe one article or book chapter per year, or maybe one every several years. The rest of the civilian faculty—the "old school," left over from earlier years—seemed to me to be dead intellectually. They never wrote or published anything. Meanwhile, on the military side of NWC and among the brass, there was no interest whatsoever in research and publication, and a positive disdain for it. If you're a military officer, research and writing carry no rewards.

I found this disdain present and prevalent in almost everything I did that was connected with research:

- In my first year at NWC, taking the Title 10 position seriously along with the renewed emphasis on research and publication that I'd been assured of verbally, I continued a vigorous writing agenda. But I heard the feedback quickly from deans and colleagues that people resented my productivity, were upset that I produced more than they did (which was usually nothing), that I had more opportunities than they to give outside lectures and contribute to anthologies, and that I did my research and writing on government time. To me, this sounded like the ancient problem of the Russian peasantry: if one peasant rose above the rest, the others would conspire to bring him down. These complaints came mainly from faculty who were pre-Title 10. Already in those first months at NWC, my journal notes (Vol. L, p. 102) record this reaction: "This is really a teaching institution where research is thought of as peripheral."
- Dale Herspring, a new Title 10 hire and an expert on both the East German and Russian armies, comes to my office to tell me he's exploring other job possibilities and will be leaving NWC. The official reason for his leaving is lack of tenure; the real reason is the absence of appreciation for research and the College's lack of understanding of the relationship between research and teaching.
- Why so little appreciation for research? We're told by the XO that the new commandant is opposed because that introduces "inequalities" into the College. Research and publication by the civilian Title 10s make the other half of the faculty, the officers in the Department of Military Strategy and Operations, look bad by comparison since they do *no* research or writing. Therefore, in the name of preserving harmony, no one should do research. What strange logic this is!
- The other College administrators, all military officers, similarly have no sense of the research mission or its relation to teaching. For them, research is below community service in their list of priorities. Research is something you do "for yourself," not because it has any value or relation to teaching.
- Meanwhile, we hear that other faculty members are complaining about the research Herspring, I, and a handful of others do because it puts added

demands on the secretaries for typing manuscripts and on such NWC services as copying and mailing. Of course, it is the nonproductive faculty who are doing the complaining. As a consequence, I receive a visit from my department head saying my research efforts are "non-curricular" and, therefore, must be done in my own time—at night.

- The pressure mounts. Heretofore, my department head, while seeing no value in my research, had at least tolerated it good naturedly as something "Howard does." But now, acting under orders from above, the prohibition comes down: no secretary is allowed to work on my research manuscripts; the mail system cannot be used for "personal matters" like research; and no research or writing can be done on government time. I appeal and actually win some victories, including typing time and mailing services. When I'm not teaching and have no other obligations, I continue to work on my research and writing during regular work hours. Meanwhile, the resentments on the part of the nonproductive faculty continue to fester. Whoever heard of a would-be academic or teaching institution that actually banned research?
- I receive another communication from my department head. My research work is now designated as a "spare-time activity," to be undertaken on non-government time. I'm told that my research work may "detract" from my teaching instead of, as academics believe, adding value to it. In a visit to my office, the department head tells me—quite sheepishly; it's obvious he's uncomfortable conveying these command decisions—that my research work is a burden on the secretaries and NWC facilities. He tells me I'm viewed by the command as having an "attitude problem" and not demonstrating "collegiality." Presumably, collegiality would, therefore, be restored if we did no research whatsoever and performed at the level of the lowest common denominator. I also inquire of the head how my research is hurting my teaching if my teaching evaluations are the highest in the College. No answer is forthcoming.
- When a new person, himself a civilian academic who, therefore, should know better, becomes department head, the situation becomes worse. There is even less emphasis on research. The head tells us that scholarship is "something you do on your own"; it's of no value to the College. How sad! How shortsighted! NWC has no sense of the prestige that accrues with having a faculty on the forefront of innovative research or of the connection between research and teaching. I get the feeling the NWC command is fearful that research by the faculty may lead to conclusions contrary to DOD policy, and that will get the College in trouble with the Joint Chiefs and jeopardize the career plans of NDU leaders.
- NWC issued new rules on typing. There are only two categories of typing: official (government, NDU) and private—what you do for *and by* yourself.

There are no in-between categories, such as "research" or "professional." A professional or research function independent from NWC is considered part of your "private life." That can't be done on NWC time or using its facilities. Earlier deans and department heads have generally been relaxed about enforcing these rules, but not the new head. The department secretary told me she had been instructed by the head not to do any letters, memos, let alone manuscripts unless they were directly government-related. I thought this was petty and we argued with the dean and went back and forth on this issue during my entire time at NWC.

- The commandant issues a memo on priorities. There is no prior consultation with the faculty; the memo comes down as a decree-law. All the emphasis is on teaching, course design, curriculum development, advising and counseling, and service to the College. Scholarship and professional development are near the bottom of the list, trailed only by Outreach. Scholarship and research are portrayed in the memo as "extra-curricular," done "when time permits." With this memo, I start making plans to go back to Amherst.
- A new commandant is assigned to the College. He spends the summer reading the student evaluations and pouring over the curriculum. He reports to us that he's concluded NWC is "not broken"—well, thanks a lot!—and he's reported that conclusion to the Joint Chiefs and its chair, General Shalikashvili. He tells us General "Shali" agrees with him that we are a *teaching* institution; scholarship has a lower priority. Moreover, he tells us our purpose is to teach U.S. national security doctrine—from "the book," no questions asked. It's a deductive form of learning: you start with "truth"—U.S. security doctrine—you learn that by heart, and you deduce from that operating principles for application in specific circumstances. I'm finally getting it—research by the faculty might contradict the official national security doctrine, and that is impermissible.
- Aha, lots of things now fall into place! When I'd first gone to NWC, I'd conceived a project and gotten a contract for a book on "U.S. Foreign and Strategic Policy in the Post-Cold War Era." I thought, given the expertise assembled there, that NWC would be the ideal place to carry out such a policy assessment, and, furthermore, that my NWC colleagues, or at least some of them who still had scholarly abilities, could be enlisted as contributors. But the project never received any support from NWC; the command and the Joint Chiefs sought to put barriers in our way; and the faculty colleagues not chosen to participate sought to undermine the entire project. The book was eventually published[4] and is, in my view, a first-rate book and certainly a contribution to the post-Cold War debate over foreign policy. But it does not correspond to official security doctrine at the time and, therefore, I can

see why the brass would be upset. The project was a learning process for me in more ways than one.

- A week after the earlier memo on priorities, we receive a new missive from the commandant. Its tone and contents are completely different from the previous one: it's actually a polite, nice memo—no huffing, puffing, and pulling of rank. The memo restores research and service to a higher priority. We're all dumbfounded. I had met with the dean to object to the earlier memo and senior scholar Paul Godwin, reflecting faculty opinion, had also weighed in with a memo to the commandant. So the faculty has some influence after all? We're amazed! But where now do we stand, with such contradictory messages?
- The answer is not long in coming. Nothing has changed! Teaching is king; research is denigrated. The dean pulls me aside to explain. It's *not just* that what we write might violate DOD policy or doctrine. It's also that some faculty (me!) write more and are more productive than others. And that introduces inequalities into the faculty's ranks that are disruptive of—the dean's words—"unit morale." But what kind of a "unit" are we civilian professors at NWC? It's not as though we're all in a foxhole together where survival depends on unit morale. No, we're in peacetime; the Cold War is over; we have no enemies; and besides, this is a "college," not a foxhole. It's the application of a military principle—unit morale—to an institution and circumstances where it's completely irrelevant. How strange! But at least now I understand.
- One of our colleagues, Dick Melanson, has written an excellent book on "Reforging [Foreign Policy] Consensus." We want to use it in our classes but are not permitted to do so, the NWC legal office rules, since that would constitute conflict of interest because of the royalties accruing to the author. That's ridiculous; all of us in this business know we don't get rich from the meager royalties our textbooks earn. The prohibition against using Melanson's text again shows how little NDU officials know about the teaching or research endeavors.
- We now receive a memo from the NDU president on how we are to spend our time. The answer: *in the College* and *at our desks*. The memo goes on to say that, if we work at home, we can *only* do classroom-related reading and research. That's ridiculous: how can it be enforced? Will they spy on us at home as well? If we go out-of-town, the president's memo continues, we must apply for leave. Otherwise, we are assumed to be "on duty" and "at our stations." The language is, of course, all military and entirely inappropriate in an academic setting. It's another application of the military mind and culture to an academic institution where it does not fit. The memo is accompanied by a warning that, if we travel without leave

or are not at our "stations" when an accident occurs, our insurance may not cover us.

- Equally worrisome is the case of my colleague, Jed Snyder. Jed worked at the Institute for National Strategic Studies (INSS), our sister institution on the NDU campus. Jed was suspended without pay for doing private research work, for remuneration, while on NDU time. In academia, of course, we do this all the time, frequently supplementing our salaries by our writing, lecturing, and consulting. But not at NDU or in the federal government generally: your time belongs *exclusively* to the government. No sidebar activities; no extra compensation from private sources.
- Wow! I'm very vulnerable on precisely these same grounds. In my academic life I seldom distinguished, and certainly not sharply, between the academic writing, lecturing, and consulting that I did, *and* the more popular writing and lecturing that went into my textbooks or books written for a general reading audience. Of course, *I* knew the difference, but my view was that these materials all came out of my research and my general storehouse of knowledge and information. And at NWC I had continued that tradition without giving it a second thought: I wrote textbooks as well as more academic works and I continued my lecturing and consulting work built up over many years, some of it done on College time. But now, unlike in academia, I'd have to be more careful, cover my trail, and be prepared to answer if ever confronted over these extracurricular activities. The idea that I'd actually give up these activities to conform to NDU rules, however, never crossed my mind. I'd go back to Amherst before I'd allow that to happen.

TENURE

For faculty at civilian colleges and universities, tenure is an essential component of the job. The rules in most civilian colleges and universities say that, after a five- or six-year apprenticeship as a teacher and scholar, you're eligible to be considered for tenure. You may also be eligible, if you're a young scholar, for promotion to associate professor at the same time.

To receive tenure and promotion, which amounts to a permanent, lifetime contract, you must demonstrate to your peers a commendable level of research productivity, publications, and teaching accomplishments. Of course, you can still be let go for malfeasance, unexpected financial exigencies at the college, or "moral turpitude" (at some colleges, which may include drug use or sexual harassment) or "cause" (which covers a range of exigencies). And in recent years many universities have instituted a post-tenure "seven-year review"—i.e., even with tenure, you still have to periodically demonstrate continued productivity. So tenure does not quite assure you lifetime employment free from all future obligations. What it does do is give you greater security in your position, the ability to stand up to your superiors (department heads, deans, provosts) if needs be without fear of being fired, and the protection of free speech even if your ideas are unpopular.

The institution of tenure runs contrary to everything the military mind believes in. It is completely foreign to the military way of thinking. The military promotion system is one of up-or-out. Either you punch the right tickets, serve in the right billets, and thus get promoted to the next level—or you are out. At every level there are fewer and fewer slots available and the candidate pool gets progressively smaller and more talented. There is no guarantee of permanent employment; either you make the next rank or your career is finished and you look for other employment. By the time you get to the level of lieutenant colonel, colonel, and then "flag officer" (general or admiral), the number of persons at those ranks is very small indeed.

And even then, if you wish to make it to two-star, three-star, or eventually four-star level, you have to keep delivering in order to avoid being weeded out—denied promotion. Hence, the idea that after a "mere" six or seven years an academic can be given what amounts to a permanent or "lifetime" position (tenure) without being required to continue delivering runs counter to everything the military command, especially at the highest levels—including the president and commandant at NDU who've successfully gone through this demanding, lifelong gauntlet—believe in. So the stage is set again, over this issue, for another conflict between the military and the civilian cultures.[5]

When I first arrived at NWC in the late summer of 1991, I immediately sought out some of the faculty who'd been around for years and who could teach me about the institution. One of these was Al Pierce who'd worked for Robert Wood in Boston when Wood was president of my former university, the University of Massachusetts. Al was a shrewd and perceptive observer; tenure was only one of many topics that we discussed. Al told me that the military was very worried about the faculty's rising demand (including from me) for tenure. First, he said, as in the military itself, they want all "the parts" (the faculty) to be interchangeable. That leaves no room for some faculty to be singled out as especially deserving. Second, he informed me that commandants and NDU presidents frequently come in with their own list of favorite faculty whom they want to see appointed at the College. It is in the nature of their "commands" to be able to do that, and tenure or permanency for the faculty would prevent them from doing so.

Third, Pierce warned, the military is very worried about "deadwood": if you give a faculty person tenure, what prevents that person from "going to sleep" and never being productive again? The answer is: even with tenure, you need to keep on producing. And to ensure that, universities have those seven-year reviews mentioned earlier. Though the faculty at NWC would continue to discuss the issue, my lunch with Al Pierce was particularly useful in pointing out the roadblocks that might lie ahead.

The tenure issue continued to simmer at NWC, mainly on the back burner. I continued to talk about it openly but found many of my colleagues, including some of the senior civilian faculty, unwilling to discuss it for fear of riling the commandant and the command structure. Finally, at the end of my first year at NWC, Paul Godwin, a civilian faculty member and then serving as an associate dean, came to see me. Paul's message was that we should not confront the command on this issue, and certainly not at the upcoming faculty retreat. That would only get the command's backs up, he said, and provoke all-out resistance; better to do it quietly. Do it, he said, as we had recently done with the issue of sabbatical leaves: set up a committee; see how it's done at other PME schools; then present our recommendations quietly and privately to the commandant. For, as Paul insisted, having much more government experience than I, all it takes is a phrase inserted between two commas in either the legislation or the regulations ("the regs") for tenure to come to NWC. That is how it came or is coming to the other PME schools, he insisted; there's no reason it can't also come here.

The next time tenure came up for serious discussion was at the NWC retreat in Colonial Williamsburg in August 1992. At precisely the venue at which Paul Godwin had warned us not to bring the issue up. But I was not the one who brought it up; tenure was on everyone's agenda including the command.

Each year the College would find a nice resort area like this, and the entire faculty, military as well as civilian plus the command, would go off for two to three days of discussion and "bonding." Tenure was an ongoing issue of concern; I learned a lot from the discussion. It turned out that faculty colleagues, Terry Deibel and Bard O'Neill, hired before me and the other Title 10s, had "permanent status" as "Title 5s." That meant that, if they were ever let go at NDU, the federal government would have to find another comparable job for them in the federal service—permanent tenure in all but name.

But we also learned at our retreat that the Joint Chiefs, NDU's president, and our own commandant (all products of the military up-or-out system) remained unalterably opposed to tenure. For one thing, at our sister institution (ICAF), in the recent past, there had been a scandal with an incompetent faculty member who, because he was tenured, proved almost impossible to get rid of. For another, the point was made by an NDU lawyer that Title 10s, because they are federal budget-related (which means the funding could go up or down), would seem not to permit tenure because that implies a permanence that, in fact, budgetarily, does not exist. Title 10s provide for *funding,* we were told, not necessarily for guaranteed *positions.* Third, we were told by the NWC command, it wants faculty "turnover" so that fresh knowledge on new, hot issues was constantly being fed into the College. Nevertheless, the command told us, opening the door to future discussion, tenure might not be incompatible with a good, dynamic faculty.

Shortly after the retreat, a "Committee on Faculty Appointments" was established to look into tenure and other issues. With my interest, now quite vocal, in the tenure issue, I was asked to be a member of the Committee, along with Colonel Craig Ham, chair, Colonel Bill Drennan, and Professor Ron Tammen. We met in mid-September 1992 and occasionally after that; it was a hard-working committee and I learned a lot from the proceedings. We investigated, for example, the situation of tenure in civilian universities as well as other PME schools and found that, in places like West Point, the Naval Academy, the Naval War College, the Air Force Academy, the Air War College, and the Navy Post-Graduate School, there was a clear trend toward permanent or what we called "indefinite appointments." The fact that many of our Title 10 appointments were based on the federal budget and, therefore, could be subject to change on a yearly basis made the granting of tenure per se difficult, but practically the same thing could be accomplished by granting indefinite appointments or through a conversion of some of our appointments to what were called "Title Five Competitive," which, as in the cases of Terry Deibel and Bard O'Neill, *guaranteed* them a comparable position within the federal government even if something bad happened at NWC.

The report our committee wrote for NWC, fifteen single-spaced pages, was, if I do say so myself, a very thorough, persuasive, and subtle document. In writing the report we were guided by "old hand," Paul Godwin, who served as an "ex officio" member of the committee and who was writing a similar report to ours at the NDU level. So in our report we did not "take on" the command structure, which we knew would be opposed to any pro-tenure recommendations. Instead, we were very deferential to the command, marshaled our evidence from the other PME schools very carefully, presented the arguments strongly but without embellishments, showed how "indefinite appointments" would benefit the College and its recruiting, indicated that this would be only a minor "administrative change" rather than a revolution in how NWC was run, and stressed that, even with these changes, the command structure would not be altered and that faculty could still be removed for a variety of causes. We must have done our jobs well because, in commenting on the report, all the deans and department heads took pains to lavish praise on it.

In working on this report, I learned a lot, not just about the other PME schools but also how NWC worked and the sensibilities of its command system. I learned, for example, that, even if we succeeded in establishing a form of "tenure" at NWC, we would still be less protected in our jobs than at a civilian university. Tenured professors can also be let go "for cause" or for "financial exigencies" at civilian universities, but at NWC you could be let go if the mission changed, if the workload or organization of the College changed, if you were not reappointed, or for "other similar or compelling reasons." The language was so vague that, if they wished, the command could still get rid of people for almost any reason. But it would not be without considerable conflict and cost to the College; we on the Committee considered our report a major breakthrough.

In the following months our committee report was discussed thoroughly, by both the faculty and the command. Memos went back and forth from deans, department heads, the commandant, and individual faculty members. All praised the thoroughness of the report and its recommendations while also, on occasion, offering an amendment or clarification. We on the Committee and among the civilian faculty had high hopes that our recommendations would be enacted.[6]

But then a number of things happened, outside of our control, that dashed our hopes. First, with the end of the Cold War, the Clinton Administration in 1993–94 began to drastically reduce the Defense Department budget. This was the period, after the collapse of the Soviet Union but before 9/11 and the war on terrorism, of the so-called "peace dividend" when both the Administration and Congress, eager to get their hands on the money for their own pet projects, sought to dramatically reduce the military budget. At NWC we were facing a 15 percent budget cut that included cutbacks in supplies, staff, faculty, and even the possible cancellation of the springtime "Regional Studies" trips. How could we argue for permanence or "tenure" in our positions when other staff were being let go and programs were being curtailed?

Second, the command structure at NDU, including two successive presidents, came out strongly against the tenure proposal. It's quite likely that our own commandant might have gone along with our proposals; but when his superiors opposed them, he had no choice but to oppose them also. Not only did the NDU presidents veto our tenure and reform proposals, they also in succeeding years centralized virtually all decision-making at NDU in their own hands, taking it away from the College and lessening its autonomy and concentrating almost all power at NDU president and staff levels. All this was done, in my view, to the detriment of the National War College—it faculty, students, and mission.

The third reason the tenure issue got sidetracked, at least in my mind, had to do with my own individual case. By this point in the mid-'90s, I had become thoroughly unhappy with the authoritarian and oppressive situation at NWC, including both the dean, and the department head, as well as the commandant. I was working on taking a leave from NWC so that I could teach a course at the University of Massachusetts or, possibly, leaving NWC altogether for Amherst's greener pastures. In addition, this was the period when UMass was in the process of establishing an endowed chair for me, an especially attractive prospect that, when it happened, led me to abandon all interest in reform at NWC and to scurry back to my wonderful, *tenured* position there.

As these ingredients were all stirring in the pot of my own decision-making, I made an appointment to see the NWC commandant to see if we could arrive at a resolution. To my request, first, for a leave without pay so I could teach for a year back at UMass and thus preserve my professorship there, the commandant was surprisingly cool. Even though it would not cost him anything, he said that, if he let me go on leave, he was afraid in the current tight budget situation that he would lose my position permanently—even though he eventually gave me the leave.

Second, I explored with him my possibilities of getting an indefinite appointment even if that was not yet College or university policy. My thinking was, if I could get a five- or six-year appointment at NWC instead of just the current short-term contract, I would consider that the near-equivalent of tenure and would thus plan to stay on at the College. But the commandant argued there were no procedures in place to decide whether I could get an indefinite appointment, nor would there be any by the time I needed to decide. This was nonsense: he could have relied on the recommendation of my dean and department head; he could have set up a joint faculty-administrative committee to decide; or he could have made the decision himself as commandant. That's, after all, what commandants had always told us in the past, that their role was to *command.*

We already know the ending of this saga. In the face of the increasingly unhappy atmosphere at NWC and the unwillingness of the command structure to bend at all, I decided to go back to my civilian university, UMass. The issue was not just the "push" at NWC but mainly the "pull" at UMass, which now endowed a chair for me and treated me very generously. Meanwhile, the "tenure" or "indefinite appointments" issue at NWC languished. It had been beaten back and vetoed by two successive NDU presidents. Over time, NWC would see the value of doing things differently, but certainly not in the short term.

There were many such decisions, like that on tenure, at NWC. Over time I came to see that, not only was NWC not going to become, in my time there, like a civilian university, but that it would *never* become that. And furthermore, that the command structure there and its arbitrary and capricious decisions would also prevent it from becoming even a decent graduate program. It really was time to leave.

NOTES

1. Technically, faculty writings are only reviewed by the academic dean to ensure that they do not use classified materials or violate security matters. But many of my colleagues discovered that, if they deviated from DOD positions, they could quickly be in bad trouble with the command structure at NWC.

2. During the past decade, the College has instituted a real sabbatical program that results in two or three faculty members being on paid leave each year.

3. In the Advanced Seminar I taught on Latin America, I used my own (best-selling, seven-edition) textbook, *Latin American Politics and Development* (Boulder, CO: Westview Press, 1985), and I remember how surprised—and pleased—my officers were that their instructor had written *the* text in the field. Similarly, in the Core Course: occasionally my writings and ideas would show up in the readings and syllabus. The students were surprised and pleased when that happened, but the command was dumbfounded and didn't know how to react.

4. Howard J. Wiarda (ed.), *U.S. Foreign and Strategic Policy in the Post-Cold War Era* (Westport, CT: Praeger Publisher, 1996).

5. Actually the differences between these two systems may not be as wide as it's sometimes portrayed. On the civilian academic side, the struggle in graduate school, passing comprehensive exams, writing a Ph.D. dissertation, finding a job, teaching, publishing, getting tenure, etc. is probably just as demanding, if not more so, as passing through the junior ranks of the military. And then ever afterwards you have to go through those seven-year reviews which are not all that different from the passage from one rank to the next in the armed forces. On the military side, if you're an officer, say, at the lieutenant colonel level, you pretty much know what the promotion rate is to the next or full colonel level—about 50 percent, depending on needs and the year, which is not all that different from the percentage who achieve tenure. And, of course, in wartime or when the nation is threatened by terrorism, as at present, the promotion rate goes way up—so much so that in certain ranks and billets it's all but guaranteed. So maybe the differences between the civilian and military careers is not all as stark as it appears.

6. In my files I have memos on the tenure and other issues from Terry Deibel, Dean Robert Hughes, Al Pierce, and former Dean Roy Stafford.

Chapter 8

Conclusion

In 1991 I had followed my wife to Washington, D.C. She had, a year-and-a-half earlier, taken a prestigious specialist position at the Library of Congress; I had agreed to stay back in Amherst so our son Jonathan could finish high school with his class. But once Jonathan graduated and had been accepted as a freshman at the University of Michigan (also my alma mater), I felt free to join Iêda in Washington. We already owned a house there, and we'd lived in Washington for most of the 1980s, so we already knew and loved the city.

In the spring of 1991 I had sent out my resume and applied for positions at several universities in the Washington area. I wrote to American University, Georgetown, George Mason University, George Washington University, and the University of Maryland. At both AU and GMU I was a finalist for the position of department chair, but neither position in the end worked out. From the other three universities, I received back very nice letters saying, essentially, "thank you for your interest. Your credentials are strong, but we presently have no faculty slots available in your area of specialization."

The prospects were starting to look glum. But then a colleague sent me the announcement of a position open at the National War College. The position fitted me perfectly. NWC was looking for a specialist in Latin American affairs who could also teach foreign policy and international relations, and who had Washington and policy experience. Well, that description fitted me to a "T." I applied and got the job. The dean later told me that my application materials were several notches above any others that they had received. NWC was then, under the Title 10 legislation, seeking to upgrade its faculty. They were very happy to have me as a new faculty member. Meanwhile, playing it safe, I had taken a leave of absence, rather than resigning, from my civilian university, the University of Massachusetts, just in case NWC didn't work out. It turned out to be one of the best decisions I'd ever made.

At first, I was thrilled to be at the War College. The dean had reserved for me one of the nicer faculty offices at NWC, overlooking the Anacostia River. The grounds and buildings at the College were beautifully maintained; everything was clean, polished, and military-spiffy. The golf course was just outside my door; the tennis courts were only a stone's throw away; and the Officers' Club provided elegant and subsidized dining. The setting, at the confluence of the Potomac and the Anacostia Rivers, was beautiful; and the location, only a mile from Capitol Hill and about the same distance to the White House, K Street, and the center of the city (and all the think tanks, law firms, interest groups, and restaurants of the area), was perfect. I had free parking (a big consideration in downtown Washington), the security of being inside a military base, and they paid extremely well. What could be better than that?

I do not have a military background. My father had served in World War I and three uncles were in World War II, but they had mustered out subsequently and had not followed military careers. In my own case, I was too young (twelve) for the Korean War and too old (past twenty-six), married, with a child, and in an occupation, teaching, that made me ineligible for Vietnam. I hadn't grown up in a military family or ever contemplated a military career; indeed, like most Americans, I'd never had much contact with the armed forces. I'd worked for the State Department earlier in my career but never for the Defense Department, although I had previously done some occasional lecturing and consulting for DOD, and I'd briefed some of its officers on Central America.

On the other hand, I did not initially think of my War College appointment as a bad fit. Unlike many academics and persons of my generation, I was certainly not anti-military. My 1970s experience with the Committee on the Present Danger and my 1980s years at the American Enterprise Institute had led me to be sympathetic to U.S. defense needs. My son-in-law was a Naval Academy graduate and a Marine officer, and I was sympathetic to the discipline and order required in a military career. I was a Midwesterner and a (liberal) Republican and, therefore, by background and culture, not all that far at variance from the military officers I'd be dealing with. I view myself as flexible, adaptable, and able to adjust to new situations—including that of the War College. If any civilian academic could fit in here, I thought I could. To be truthful, however, I think I expected the War College and the National Defense University to be just like other academic institutions and universities with which I'd been affiliated, and they assuredly were not that.

Right from the beginning, I sensed and saw intimations of trouble ahead. It was immediately apparent that the War College structure and command system were much more top-down and authoritarian than I'd thought. It was made very clear to all the new faculty that this was, first and foremost, a *military* base, and an *Army* base at that, and not at all like a civilian university. Already in the first two weeks, although there was nothing about this in my contract, I had been "tasked" to lecture to a Capstone wives' group and to one of the Reserve Component National Security Course (RCNSC) programs, without any "by-your-leave" let alone extra compensation. Eyebrows were already being raised about my research and writing agenda, which continued unabated at NWC. My journal notes show that by early September, only a couple of weeks after the course of study at NWC had begun, I was already recording that I was very happy I'd hung onto my tenured full professorship at UMass. I could already see there would be difficult times ahead at NWC and that I might want to go back to my university teaching position.

I went into this job at the War College as a person very sympathetic to the U.S. military and to what the War College was seeking to do. And that is, as I perceived it, to teach these budding lieutenant colonels and colonels about foreign policy, international relations, and the countries and regions—what we in political science call "comparative politics"—with which they would have to deal as high-level ("flag officer") military and national security officials. That, *of course,* in my view, would involve original research and writing on my and other faculty's part, and the conveying of that knowledge to these future "Brent Scowcrofts" and "Colin Powells"—i.e., to those who would in the future man key Pentagon, National Security Council, and other strategic positions. There is no question in my mind whatsoever that is what teaching and the curriculum at NWC *ought* to be.

How naïve I was in those days! I really believed—and still do believe—that was my charge as well as the College's role. I guess I'm an idealist in this respect. Or maybe this harked back to my Calvinist background. I saw it as my duty and obligation to educate these officers on what I knew and to convey to them whatever specialized knowledge I had.

I should have known better. Within a matter of days, certainly weeks, I saw this was not going to work the way I thought it would. Here are the main problems as they began to impinge on my work and thinking:

1. The War College was extremely authoritarian and top-down. It did not at all function like any college (lower case) I'd ever seen. It was not a "college" or "university" at all but a military base run on a command system. It had not made the compromises necessary to be both a military institution *and* a serious teaching and educational institution.
2. The curriculum was based on a set of prescribed syllabi and texts. You were given a lesson plan and a set of readings, and that was pretty much

it. No deviation. You taught "from the book." True, the curriculum was revised from year to year but only in modest ways; the basic lesson plan was pretty well set. It was more like a manual in a technical school than a serious graduate curriculum. That kind of official, prescribed curriculum may be defensible in lower-level schools, but it has no place in a supposedly MA-level university program.

3. With a set curriculum like this, there was no way to incorporate the faculty's original research and ideas into the curriculum. The course syllabi prescribed in great detail how and what you taught. That was it! There was no room for new or original ideas.
4. It quickly became clear that NWC was a teaching institution—and only that. With a prescribed curriculum, no value would be placed by the College on research. The *entire focus* was teaching, not research. This is like elementary school: the "school board" (the NWC command) chooses the curriculum and textbooks, and the teachers teach *only* from that. Hence, the actions taken by NWC to limit or eliminate altogether my research and writing work.
5. The College had only a one-way flow of communication. It all came from the top down (the command system); there was no institutionalized way (faculty council, faculty senate, certainly no union) for the faculty to make its voice heard at high college and university levels. After all, it was the faculty that taught these students, not the command. We should have input.
6. The College put the students on a pedestal. Above the faculty. This inversion of roles was purely political: it had to do chiefly with the fear the command had that, if it didn't treat the students well, they would either report that to the Joint Chiefs or else, when they became generals and admirals, they would wreak their own (budgetary or otherwise) vengeance on the College. Either way, this chicken-hearted policy would be bad for NWC.
7. The College proved unwilling to make its compromises with its civilian faculty and academic side. It functioned entirely as a military base and institution. Not only was there no support for research, there was no regular, institutionalized system of tenure, sabbaticals, promotions, or faculty merit review.
8. I don't think the military brass who run NWC (we take our orders from the Army and the Joint Chiefs) have the foggiest notion what a college or university is all about. They think of the College as being like a "trade school": you teach from "the manual"—i.e., the prescribed curriculum—and the students absorb the key lessons. But at the graduate level you can't do that: you need new ideas; you require academic freedom; you

must value original research; and you have to allow for new concepts and subject areas—for example, irregular or asymmetrical warfare, peace-keeping, nation-building, democratization—to be introduced into the curriculum. Look what happens when you fail to do that: you get disasters like Iraq, Vietnam, or Afghanistan.

9. Over time, I also came to see that all this authoritarianism at the College, this top-down system, and the ignoring of and disdain for the civilian faculty and its research were all part of a gigantic military inferiority complex. It almost sounds, without being egotistical about it, like General Lesley Grove trying to manage the famed Manhattan Project. Essentially, the military hierarchy does not treat the civilian faculty with respect, distrusts "intellectuals" in general, suspects the faculty is on the left politically, and anti-war or anti-military. Therefore, it needs at all costs to rein it in, patronize it, but also keep it under control.

Shortly after my arrival at NWC, my journals record that I saw NWC as composed of two completely different cultures and ways of doing things. The military culture tended to be Southern and Midwestern, Christian, Republican, and rural or small town. It was also derived from the military institution itself and its culture and structure: top-down, hierarchical, disciplined, orderly, everything-in-its-place. The military institution called for orders to be carried out unquestionably, for the mission to be carried out at all costs, and for "the troops"—including the civilian faculty—to "stand up and salute"—i.e., to concur in all policy decisions. When we were "tasked" to do something, in the military culture that meant that you did it, happily and without dissent.

All this is quite at variance with the "academic mind" or the academic "political culture." That culture, in sharp contrast to the military one, tends to be East Coast, secular, liberal-Democratic, and big-city. It is also, above all, open-minded, free-wheeling, questioning of authority, questioning of any and all established truths, and unwilling to take orders from anyone and anything. That is why the civilian faculty *insists* on free speech, academic freedom as an absolute, tenure, sabbaticals, time and support for research, and the ability to follow one's scholarly and research agenda wherever it may lead and whether it is in accord with DOD policy or not. As is readily apparent, the academic political culture is diametrically opposed to, or at the opposite end of the spectrum from, the military one.

How has the War College handled this problem, this conundrum, this contradiction? The answer, which may come as a surprise to some, given the conflict-prone way we have posed the question, is: in the past, not altogether badly. When NWC long-timer Roy Stafford was dean, for example, and he is a man (one of the few) who could genuinely bridge the academic and the military cultures, he was exceedingly careful in selecting persons for the NWC faculty. For the civilian faculty, Roy was careful to choose only those academics who had prior governmental or policy experience—i.e., those who were used to operating in a bureaucratic context and didn't get their backs up when given an order. Who knew how to compromise, in other words. Even in my case, coming directly out of academia, I was told once that I was selected more for my Washington, think tank, and policy experience than for my long list of scholarly publications.

On the military side, Roy tried to do the same thing. He tried to get military faculty who had advanced degrees and/or prior teaching (at the service academes or elsewhere) experience—i.e., who had some academic credentials themselves. After all, it would not do, à la my first commandant ("Every military officer is a teacher."), to stick any old tank driver, boat captain, or fighter pilot into a classroom and expect them, at this level, to succeed. When Roy received a name from the services and the Joint Chiefs of an officer who was unacceptable by these criteria, it was not unheard of for him, a "mere" colonel, to send it back and request another nomination. But over time two things happened: (1) Roy lost his ability to influence the selection process this way, and (2) future deans were not as careful and meticulous as Roy had been in selecting faculty. And that has led to many of the problems and conflicts described in this book. In fact, since my time at the War College, the situation of the military faculty has only gotten worse: almost no Ph.D.s, very little if any teaching experience, and fewer qualifications for the position. This decline in the military faculty was accompanied by a succession of commandants with no teaching or educational experience, a fundamental distrust of the civilian faculty, and no intellectual curiosity.

Let us take for granted, as I do, that there will *always* be inherent tensions and simmering conflicts between the military and the civilian academic cultures at NWC and the other PME schools. Such conflicts are "in the nature of things"—i.e., inevitable in any institution that brings together two such contrasting groups of people. Given that, what can be done about it? I see three possibilities:

1. Do nothing; maintain the status quo. That is, keep NWC as a military base, run on a military chain-of-command basis, issuing orders, allowing no or minimum faculty input, running roughshod over civilians and civilian faculty preferences and ways of doing things. I don't see that as a viable option: (a) it makes for a very bad "college" or "university"; (b) it risks an embarrassing and destructive explosion or blow-up at some point between the two cultures; and (c) it makes it almost impossible to recruit a top-flight civilian faculty at NWC.

2. Abolish NWC and NDU. Although I have concluded this is not today a realistic option, it *was* considered seriously in the past. The argument is: why have NWC at all when a considerable variety of good civilian graduate schools could provide the same, or better, level of education in foreign policy, comparative politics, international relations, and security affairs? Why couldn't the armed forces, as well as the State Department, CIA, and other agencies, just "farm out" those selected and allow them to pursue a Master's (or Ph.D.-level!) program at a good civilian university? Among other things, the costs would be far less.

   That option was treated seriously when NWC was first established in the '40s and at several (usually crisis) points since then. But I believe it is now the established and even settled view within the Defense Department, and in my own mind as well, that in its blend of foreign policy studies and the use of military force, "soft power" and "hard power," there is such uniqueness in the institution of NWC and its overall program that it ought to be kept as a separate training facility. I can't imagine a civilian university program doing quite what NWC does in training senior-level military officers for high-level posts, especially now given the decline of university-based security studies programs since the Cold War, or putting together a package of offerings that matches what the Defense Department needs.
3. That leaves the third option: reform the system. And bear in mind that I'm talking about a wholesale reform here, not just a mere tinkering with the system. Such a reform would require a new Goldwater-Nichols Act—and more! What would be the ingredients in such a major reform effort? Here we list only some of the large reforms needed; the details will all have to be worked out carefully.
   a. Less tight control by the Joint Chiefs, the NDU president, and the NWC commandant. NDU and NWC will have to be loosened up; it cannot continue to be run strictly as a military base and with a military command system. Control over NWC and NDU may have to be transferred from the Joint Chiefs to the Office of the Secretary of Defense (OSD) or perhaps even a joint DOD-State Department-civilian academic control board.
   b. Civilianize the administrative structure at NDU and NWC. Bring in as many civilian deans, department heads, and provosts as there are military. Make the civilian provost equal to the military president or commandant. Do some hard bargaining over these respective roles and responsibilities.
   c. Bring the faculty in as co-participants in decision-making. Tap their talents and expertise. We are not "all" flag burners and anti-war rebels.

Civilian universities do this through a faculty council or senate; NWC could, to its great benefit, do something similar.

d. Remove the students from the pedestal they presently occupy. Stop cow-towing to them. Treat them like any other graduate students; after all, they're there to *learn,* not to be treated with kid gloves. Whoever heard of a college or university whose administrators were scared of their own students?
e. Bridge the civilian-military instructor gap. As Roy Stafford once did. Have entrance requirements for the students so the College helps select them, not just their own services. For the faculty, accept only those military officers who have teaching experience and the appropriate post-graduate training; accept for the civilian faculty only those who have solid academic qualifications as well as prior government/Washington policy experience.
f. Bridge the military-civilian cultural gap. Bring the "two cultures" together. It's not an impossible task. That will require a variety of changes in how NDU/NWC is run, administered, the personnel in charge, the values and operating procedures of the institution.
g. Change the curriculum in fundamental ways. Abandon the rote-learning system. Teach strategic *thinking.* Update the readings and the syllabi; make them flexible and adaptable so as to accommodate new subject areas such as a-symmetrical warfare, conflict resolution, peace-keeping, nation-building, and others. Build adaptability into the curriculum itself.
h. There must be more emphasis on original research. By both the military and the civilian faculty. And that research must be built back into the curriculum and the classroom.
i. In this way, NWC can be kept intellectually exciting and alive. These elements must be reincorporated into the learning experience. You cannot teach at this level just "from the book"; you need to involve the students in the excitement of seeing the world and U.S. policy in new ways, with new facts, from new perspectives.
j. In short, what I am advocating is a return to the original, enlightened, George Kennan conception of NWC. That is not a new, much less a radical, conception. It is only a return to basics. And that is a College that emphasizes research, contemplation, deliberation, discussion, as well as teaching as its basic functions. And all this at a high, truly post-graduate level. That is the original conception of NWC; it is a conception to which the College should return.

Index

About the Author

Howard J. Wiarda is Dean Rusk Professor of International Relations at the University of Georgia, Senior Associate at the Center for Strategic and International Studies (CSIS), and a Senior Scholar at the Woodrow Wilson International Center for Scholars in Washington, D.C. He is an expert in Latin American politics and relations and a prolific author who has published or edited over ninety books, including *Think Tanks and Foreign Policy: The Foreign Policy Research Institute and Presidential Politics*; *Harvard and the Weatherhead Center for International Affairs (WCFIA): Foreign Policy Research Center and Incubator of Presidential Advisors*; and *Conservative Brain Trust: The Rise, Fall, and Rise Again of the American Enterprise Institute.*

www.ingramcontent.com/pod-product-compliance
Ingram Content Group UK Ltd.
Pitfield, Milton Keynes, MK11 3LW, UK
UKHW012041150926
475560UK00001B/120

9 780739 150856